Gerhard Schmied

High Quality Messaging and Electronic Commerce

Springer

Berlin
Heidelberg
New York
Barcelona
Hong Kong
London
Milan
Paris
Singapore
Tokyo

Gerhard Schmied

High Quality Messaging and
and
Electronic Commerce

Technical Foundations,
Standards and Protocols

With 21 Figures

 Springer

DI Gerhard Schmied
Fa. INFONOVA GmbH
Karlauergürtel 1
A - 8020 Graz

Library of Congress Cataloging-in-Publication Data

Schmied, Gerhard, 1969 -
High quality messaging and electronic commerce: technical foundations,
standards, and protocols / Gerhard Schmied. Includes index.
ISBN-13: 978-3-642-64183-1
1. Electronic mail systems - Standards. 2. Computer network protocols.
3. Electronic mail systems - Quality control. 4. Electronic commerce -
Security measures. I. Title. TK5105.73.S35 1999
004.692-ddc21

ISBN-13: 978-3-642-64183-1   e-ISBN-13: 978-3-642-59922-4
DOI: 10.1007/978-3-642-59922-4

© Springer-Verlag Berlin Heidelberg 1999
Softcover reprint of the hardcover 1st edition 1999

Typesetting: Camera-ready copy from author
Cover-Design: MEDIO, Berlin
SPIN 10683101   62/3020 5 4 3 2 1 0   Printed on acid-free paper

Foreword

I am pleased to introduce this publication based on one of the deliverables of the NO PROBLEMS project. This is one of the projects co-financed by the European Commission in the framework of the TEN-Telecom programme.

Electronic commerce is high up on the agenda of the European Commission and is one of the priorities of the TEN-Telecom programme. Electronic commerce requires a reliable and secure communications environment. But in order to attract the large number of Europe's SMEs and the European citizens, electronic commerce support services must be widely accessible at low cost.

The NO PROBLEMS project has addressed these challenging requirements by combining two complementary technologies: X.400 messaging for the reliability and the security framework, Internet for the low cost accessibility.

NO PROBLEMS is in many regards illustrative of the TEN-Telecom objectives: it combines available technology to provide innovative services to Europe's SMEs.

This publication provides an in-depth insight into some of the key technologies expected to play a major role in the support of electronic commerce.

Robert Verrue
Director General DG XIII
European Commission

This publication is based on documents of the NO PROBLEMS project, which has been founded by the European Commission DG XIII. The project partners have been INFONOVA GmbH and DATAKOM Austria from Austria as well as ITK Telekommunikation from Germany and ATLANTIDE from France.

The project's goal was to design a trans-European messaging service where multivendor interoperability is guaranteed. In order to reach this goal only standard compliant products can be used. An additional design criterion was that especially the requirements of small and medium sized enterprises (SMEs) should be taken into account.

In recent times the importance of Electronic Commerce has significantly increased. In order to design a future oriented service that takes into account the latest technologies and trends on the marketplace, specifications for Electronic Commerce applications over the messaging backbone have additionally been produced and added to the scope of the service. They were again designed in a way that especially the requirements of SMEs were taken into account to help them to improve the way they do their business so that they can exist in the very fast changing European and worldwide market place.

This publication provides useful information of the key technologies required to implement a future oriented messaging and electronic commerce service which satisfies not only the requirements of SMEs but also those of large companies.

Peter J. Kampner
Managing Director
Value Added Network Services
DATAKOM Austria

Disclaimer

This text reflects the views of the author and does not in any way engage the European Commission.

To Mr. Joseph Bremer of the European Commission DG XIII, who was the project officer of the NO PROBLEMS Project.

To the NO PROBLEMS Team of INFONOVA GmbH consisting of Mr. Michael Blaschitz, Mr. Herwart Wermescher, Mr. Wolfgang Themessl and Mr. Thomas Kutschi .

To the NO PROBLEMS Team of DATAKOM Austria AG, consisting of Mr. Klaus Sambor, Dr. Hermann Jeram, Mr. Peter Kampner, Mr. Peter Uher, Mr. Otto Bergmann, Mr. Alfred Anders, Mr. Rene Schelbaum, Mr. Andreas Steiner, Mr. Gerhard Waska, Mr. Siegfried Knar, Mr. Christian Weigl, Mr. Peter König.

To my wife Karin, my children Claudia, Christoph, Sandra and Christina as well as to my parents and my grandmother. Very special thanks to my father who unfortunately died only a few weeks before this book was published.

Contents

Abbreviations

ADMD	Administrative Management Domain
ANSI	American National Standards Institute
ARPA	Advanced Research Project Agency
ASCII	American Standard Code for Information Interchange
ASN.1	Abstract Syntax Notation 1
ATM	Asynchronous Transfer Mode
AU	Access Unit
BRI	Basic Rate Interface
CA	Certification Authority
CAD/CAM	Computer Aided Design/Computer Aided Manufacture
CBC	Cipher Block Chaining
CCITT	International Telegraph and Telephone Consultative Committee
CEC	Center for Educational Computing
CEFIC	Chemical Industry of Central and Eastern Europe
DAP	Directory Access Protocol
DCG	Data Element Coordination Group
DDA	Domain Defined Attribute
DES	Data Encryption Standard
DES CBC mode	Data Encryption Standard in Cipher Block Chaining mode
DES EDE mode	Data Encryption Standard in Encrypted Decrypted Encrypted mode
DES ECB mode	Data Encryption Standard in Electronic Code Book mode
DIB	Directory Information Base
DISP	Directory Information Shadowing Protocol
DIT	Directory Information Tree
DL	Distribution List
DNS	Domain Name Server
DOP	Directory Operational Binding Management Protocol
DSA	Directory System Agent
DSP	Directory System Protocol
DTE	Data Terminal Equipment
DUA	Directory User Agent
DUNS number	Dun & Bradstreet Information Services number (used in commercial EDI to identify Trading Partners)
EAN	European Article Numbering

EBCDIC	Extended Binary Coded Decimal Interchange Code
EBP	Extended Body Part
EC	Electronic Commerce
EDI	Electronic Data Interchange
EDIFICE	Electronic Data Interchange for Companies with Interest in Computing and Electronics
EDI-MS	EDI-Message Store
EDI-UA	EDI-User Agent
EDIM	EDI Message
EDIME	EDI Messaging Environment
EDIMG	EDI Messaging
EDIMS	EDI Messaging System
EDIN	EDI Notification
EIT	Encoded Information Type
EMS	Express Mail Service
EoS	Element of Service
ERS	Evaluated Receipt Settlement
FTAM	File Transfer, Access and Management
FTBP	File Transfer Body Part
G3Fax	Group 3 Facsimile
G4Fax	Group 4 Facsimile
GE	General Electrics
GM	General Motors
GSM	Global System for Mobile Communication
GW	Gateway
IA5	International Alphabet 5
IDEA	International Data Encryption Algorithm
IEC	International Electrotechnical Commission
IETF	Internet Engineering Task Force
IPM	Interpersonal Messaging
IPMS	Interpersonal Messaging System
IPN	Interpersonal Notification
ISDN	Integrated Services Digital Network
ISO	International Standardisation Organisation
ISUP	ISDN User Part
ITU-T	International Telecommunication Union-Telecommunications Standardisation Sector
JPEG	Joint Photographic Experts Group
kbps	Kilo Bit per Second
LAN	Local Area Network
LANE	Local Area Network Emulation
LD-CELP	Low-Delay Code Excited Linear Prediction
LPM	Local Processing Model
MCGAM	MIXER Conformant Global Address Mapping

MD	Management Domain
MHS	Message Handling Systems
MIB	Management Information Base
MIME	Multipurpose Internet Mail Enhancements
MIXER	MIME Internet X.400 Enhanced Relay
MOSS	MIME Object Security Services
MOTIS	Message Orientated Text Interchange Systems
MPOA	Multiple Protocol over ATM
MS	Message Store
MSP	Message Security Protocol
MT	Message Transfer
MTA	Message Transfer Agent
MTL	Message Transfer Layer
MTS	Message Transfer System
MTS-APDU	Message Transfer System Application Protocol Data Unit
NHRP	Next Hop Routing Protocol
NIST	National Institute for Standards and Technology
NOPROBLEMS	Non PROprietary reliaBLe Electronic Mail System
NSA	National Security Agency
O/R	Originator/Recipient
ODA	Open Document Architecture
OSI	Open Systems Interconnection
PBX	Private Branch Exchange
PD	Physical Delivery
PDAU	Physical Delivery Access Unit
PDS	Physical Delivery System
PDU	Protocol Data Unit
PEM	Privacy Enhanced Mail
PGP	Pretty Good Privacy
PGP/MIME	Pretty Good Privacy with MIME Extensions
PKCS	Public Key Cryptographic Standard
PMBS	Packet Mode Bearer Service
PR	Per Recipient
PRMD	Private Management Domain
PSTN	Public Switched Telephone Network
PTT	Post Telephone and Telegraph Administration
RFC	Request for Comment
RFQ	Request for Quotation
ROSE	Remote Operation Service Element
RPOA	Recognised Private Operating Agency
RSA-Algorithm	Rivest, Shamir, Adleman-Algorithm
S/MIME	Secure Multipurpose Internet Mail Extensions
SMTP	Simple Mail Transport Protocol
SNMP	Simple Network Management Protocol

TA	Terminal Adapter
TCP/IP	Transmission Control Protocol/Internet Protocol
TDCC	Transportation Data Coordinating Committee
TEDIS	Trade Electronic Data Interchange Systems
TLX	Telex
TP	Trading Partner
TRADACOMS	Trading Data Communications
TTX	Teletex
UA	User Agent
UAL	User Agent Layer
UN	United Nations
UN/ECE	United Nations Economic Commission for Europe
UN/EDIFACT	United Nations Electronic Data Interchange For Administration Commerce and Trade
UNGTDI	United Nations Guidelines for Trade Data Interchange
UNTDED	United Nations Trade Data Element Directory
UNTDID	United Nations Trade Data Interchange Directory
USEC	User-based Security Model
VAN	Value Added Network
WP	Work Package

1 Introduction

The exchange of information between persons and especially between business or trading partners (TPs) is and will always be very important. There is nothing new about this statement. The only thing that has changed is the way information is exchanged. In former times this was done by sending letters and postcards. Recently information exchange by sending electronic mail is getting more important.

What are the reasons for this development? Well, there are several reasons and one main reason is certainly the enormous and rapid growth of the Internet and its standards. The first email systems were mainly proprietary and host-based ones where interoperability between different systems was not possible. Since the number of the proprietary systems increased and also the benefits of email, especially where the saving of time and money were concerned, interoperability became a very important issue.

Keeping this in mind, in the beginning of the 80s the first standards were defined such that they should guarantee interoperability between the various products available on the market. But unfortunately there was not only one standard defined for all different networks. For Transmission Control Protocol/ Internet Protocol (TCP/IP) networks the Simple Mail Transfer Protocol (SMTP) was defined in 1982 as Request for Comments (RFC) 821 and for Value Added Networks (VANs) the X.400 Series of Recommendations were defined in 1984 by the International Telegraph and Telephone Consultative Committee (CCITT). The X.400 Series of Recommendations were later adopted by the International Standards Organisation (ISO) and is the only international standard. Both of these standards were further developed and since the early 90s effort has been made to define rules to map SMTP messages to X.400 messages and vice versa.

For any future-oriented email service it will be very important to be able to deal with both of these protocols in a manner that is transparent to the user. In other words users must have the possibility to send and receive messages from both worlds without any problems.

This book includes parts of the work done in a project on behalf of the European Commission, DG XIII, called "NOn PROprietary reliaBLe Electronic Mail Systems" (NOPROBLEMS). The scope of this project was initially to design a pan-European X.400 service, but during the initial phase it was clear that Internet mail users have to be considered as well. Additionally an X.500 directory is now part of the service to improve user friendliness and to store several properties for users of the service like email addresses, postal addresses, telephone numbers, etc.

In recent times Electronic Commerce over the Internet is getting more and more important. Because of the fact that the importance of the information being sent is continuously growing, the authentication of the sender and the integrity of the data must be guaranteed. The most widely used method to guarantee this is to digitally sign the information before sending it by the use of encryption techniques. Mostly asynchronous cryptosystems are used where each user has two keys, a private one that only he knows and a public one which is known by other users of the system. To store the public keys of the users is very important and can be realised by the use of an X.500 directory as well.

The growing importance of Electronic Commerce and Electronic Data Interchange (EDI) not only over the Internet but over VANs as well is considered in the NOPROBLEMS service in the way that it provides the X.400/SMTP-MIME backbone network for sending the data.

2 The X.400 Series of Recommendations

In the year 1980 the CCITT started to develop a standard for exchanging electronic mail that relies on the OSI model. The initial version of this standard was published in 1984 and was called X.400 series of recommendations. A total of eight separate recommendations make up the basic MHS series:

- **X.400 Message Handling Systems - System Model - Services Elements**
  This part defines the message handling services that enable the users to exchange messages on a store-and-forward basis.
- **X.401 Message Handling Systems - Basic Service Elements and Optional User Facilities**
  This recommendation categorises the services into basic and optional subsets to support international interoperability.
- **X.408 Message Handling Systems - Encoded Information Type Conversion Rules**
  This recommendation specifies the algorithms that the MHS uses when converting between different types of encoded information.
- **X.409 Message Handling Systems - Presentation Transfer Syntax and Notation**
  This recommendation defines the notation and representational techniques used to specify and encode MHS protocols.
- **X.410 Message Handling Systems - Remote Operations and Reliable Transfer Server**
  Herein the building blocks of the MHS protocols and the way the seven layers of OSI protocols are used to support MHS applications are described.
- **X.411 Message Handling Systems - Message Transfer Layer**
  This recommendation specifies the message transfer services and protocols.
- **X.420 Message Handling Systems - Interpersonal Messaging User Agent Layer**
  This recommendation specifies the interpersonal message services and protocols.
- **X.430 Message Handling Systems - Access Protocol for Teletex Terminals**
  This recommendation describes how Teletex terminals access the MHS.

2.1
The X.400 Standard 1984

In October 1984 the CCITT published the initial MHS standard. It covers the base elements and core functionality for exchanging information between two users. The user can be seen as either a human being or a computer application process originating and receiving messages. In the initial version of the standard special attention was given to the exchange of messages between persons. This kind of message exchange was called Interpersonal Messaging (IPM). In the following paragraphs the most important properties of MHS 1984 are mentioned.

2.1.1
Functional Model

In the X.400 standard the MHS provides the overall message handling environment. It also provides the total support to enable the users to communicate by exchanging messages and consists of the following elements:

• User Agent (UA);
• Message Transfer System (MTS);
• Message Transfer Agent (MTA).

2.1.1.1
User Agent (UA)

It is the functional element representing the user, who is either a person or a computer application process that originates and receives messages, in the MHS model. It interacts directly with the user, performs the functions for preparing messages and submits the messages for routing. The UA also assists the user in other mailing functions, like filing, replying, retrieving and forwarding.

2.1.1.2
Message Transfer System (MTS)

It provides the means for cooperating with UAs to exchange messages. In accomplishing this, the UA and the MTS interact for the submission and delivery of messages. It consists of one or more Message Transfer Agents.

2.1.1.3
Message Transfer Agent (MTA)

It is responsible for relaying the messages to their destinations.

2.1.2
Message Structure

X.400 defines a message as an envelope containing the information for transferring the message plus a content containing the user's message. As far as message transfer is concerned the user's message content is just a byte string, which may be some standardised encoded content-type or may be designated as "undefined".

The standardised user-message content-type that is defined in the X.400 recommendations is called Interpersonal Message (IPM). In the initial version of the standard this was the only introduced content-type. An IPM consists of two primary parts: the heading containing the message control information designated for use by the office automation system, and the body containing the substance of the user message. It can consist of a number of different encoded body parts, like text, facsimile, graphics, audio or video.

Envelope + Content

Fig. 2.1. Message structure

2.1.3
Management Domains

An administration or organisation may play various rules in providing MD services. An organisation may be a company or a noncommercial organisation.

The collection consisting of at least one MTA and zero or more UAs owned by an administration or organisation is called management domain (MD). In this context, "administration" can refer either to the central PTT authority in each country or to a common carrier, known within the ITU-T as a recognised private operating agency (RPOA). Domains are responsible for naming, addressing and routing. The boundaries of domains are not restricted to the boundaries of a system of an organisation.

The MD managed by an administration is called administration management domain (ADMD). The MD managed by an organisation is called a private management domain (PRMD). The ITU-T considers that in each country the ADMDs must take responsibility for message preparation and delivery. Therefore the initial X.400 recommendations for international message transfer are restricted to ADMD-ADMD. As a result of that reaching a PRMD always requires going through the ADMD to which the PRMD is connected. This connection is established at a peer-to-peer (MTA-MTA) level. The ADMD then takes responsibility for the messages created by and destined for the PRMDs that the ADMD is servicing. Furthermore it was considered that there is only one ADMD in each country. These considerations were dropped in the later versions of the standard and now it is possible to send messages directly from one PRMD to another PRMD, and PRMDs are no longer restricted to a single country. Furthermore one PRMD can not only be connected to several ADMDs in one country, but also to several ADMDs of different countries.

2.1.4
Naming and Addressing

As a similarity to the postal service, where naming and addressing is used to clearly identify a person, the MHS recommendations use naming and addressing to unambiguously identify a UA. These identifiers are in the case of MHS 1984 called originator/recipient (O/R) names. They are added to the message envelope and used by the MTS for routing and delivering messages. O/R names are constructed from a set of attributes for flexibility. There are four categories of standard attributes:

- personal attributes
    - personal name consisting of: Surname, GivenName, Initial, Generational.

- geographical attributes
    - CountryName;
    - RegionName;
    - TownName.

- organisational attributes
    - OrganisationalName;
    - OrganisationalUnit;
    - PositionOrRole.

- architectural attributes
    - X.121 Address;

- Unique UAIdentifier (only numeric)
- ADMDName;
- PRMDName.

The MDs must ensure that each UA in the domain has at least one name. An MD does not necessarily have to utilise all attribute types in its MD when creating O/R names. There are no precise standards for how the attributes have to be used and the only mandatory elements are country name and ADMD name. Giving O/R names to special UAs is the responsibility of naming and addressing authorities. Therefore each MD has such an authority. To ensure that each MD has a unique name a national naming and addressing authority has to exist.

2.1.5
The "ITU-T Service Concept"

The ITU-T defines several types of communication utilities that are offered by public agencies to a customer as "services". Examples are telephone, telex, videotex, and facsimile services.

Another example is the MHS service which allows customers to communicate by exchanging specified types of messages electronically via store-and-forward and store-and-retrieval techniques.

The features of the MHS service are defined in X.400 and in 1984 they were called "service elements" (SEs). Features that have to be available by all MHS services are called basic service elements. Others are used at the discretion of the customer and are called optional service elements or optional user facilities. The agency offering the MHS service is required to include certain optional user facilities and can choose whether or not to include the others.

2.1.6
Message System Types and Protocols

In the 1984 series of recommendations there are three different message system types corresponding to the way the functional elements can be combined.

- UA and MTA are implemented in the same system;
- UA and MTA are implemented in physically separate systems;
- an MTA is implemented in a system without a UA (stand-alone MTA).

There are only three protocols needed in this version of the standard to interconnect all these possible types:

- the P2 protocol is the UA-to-UA protocol and consists of interpersonal messages and interpersonal notifications;

- the P1 protocol is the MTA-to-MTA protocol and comprises the specification of envelopes and delivery reports used for the main part of the message transfer functionality;
- the P3 protocol is the UA-MTA protocol and provides remote access to the MTA.

The X.400 series of recommendations define two sublayers on the Application Layer for the communication between two systems:

- the Message Transfer Layer (MTL) with the P3 protocol as submission and delivery protocol (UA-MTA) and the P1 protocol as message transfer protocol (MTA-MTA);
- the User Agent Layer (UAL) with the P2 protocol as interpersonal messaging protocol.

2.1.7
Problems of X.400 1984

The MHS 1984 standard meets the basic requirements for MHS recommendations. But there were also some additions and updates required which are listed below:

- No mailbox (message store) standard
- No distributed lists
- No full OSI stack
- No distinction between name and address
- Important service elements are not included

2.1.7.1
No Mailbox (Message Store) Standard

In the 1984 standard each UA is responsible for his own (local) mailbox. There is no standard for remote mailbox access. Furthermore it is only possible that messages are delivered to the UA in transfer priority order. This priority is determined by the sender and not by the recipient. This implicates the possibility that a message a user considers most urgent is delivered last.

2.1.7.2
No Distributed Lists

In the 1984 standard, the entire recipient list has to be known and listed on the envelope.

2.1.7.3
No Full OSI Stack

MHS 1984 was the first OSI Application Layer standard and in 1984 the Presentation Layer was not completed. Therefore it was decided to use Session Layer services directly bypassing the Presentation Layer. Later Application Layer standards were able to take advantage of the Presentation Layer and thus MHS found itself out-of-step with these standards.

2.1.7.4
No Distinction Between Name and Address

The 1984 "O/R Name" is actually an address, and there is no provision for distinguishing between an O/R name and an O/R address. This makes it difficult, or impossible, to provide services like address change and message redirection.

2.1.7.5
Important Service Elements are not Included

Functional improvements like physical delivery, security, message redirection, improved recognition and treatment of private domains, improved identification and architecture of access units to existing message services outside the MHS, and more flexible identification of body part types, were identified as absolute necessities and are therefore included in the next version of the standard.

2.2
The X.400 Standard 1988

Like all other CCITT (ITU-T) standards X.400 is revised every four years as well and as a consequence of that a new version of the standard was published in 1988. The problems and lack of functionality mentioned above led to significant changes of the initial version of X.400 in the 1988 version of the standard. The changes were not only in regard to solving the known problems and in adding new functionality, but also in the naming and in the structure of the series of recommendations.

2.2.1
New Definitions

In MHS 1988 a distinction between the MHS system and the MHS service was introduced. The definition of the MHS service is given in the F.400 series of recommendations. The definition of the MHS system is still given in the X.400 series of recommendations and it provides the technical specifications needed for the implementation of the F.400 series. Furthermore the term "Service Element" as used in MHS 1984 was changed to "Element of Service" in MHS 1988 to avoid confusion with the term "application service element" in the OSI Application Layer. A listing of the single recommendations of X.400 1988 and F.400 1988 is given bellow.

2.2.1.1
X.400 Series of Recommendations 1988

The X.400 series of recommendations 1988 consist of the following documents:

- X.400 Message Handling Systems and Service Overview;
- X.402 Message Handling Systems: Overall architecture;
- X.403 Message Handling Systems: Conformance Testing;
- X.407 Message Handling Systems: Abstract Service Definition Conventions;
- X.408 Message Handling Systems: Encoded information type conversion rules;
- X.411 Message Handling Systems: Message Transfer System: Abstract service definition and procedures;
- X.413 Message Handling Systems: Message Store: Abstract service definition;
- X.419 Message Handling Systems: Protocol specifications;
- X.420 Message Handling Systems: Interpersonal messaging system.

2.2.1.2
F.400 Series of Recommendations 1988

The F.400 series of recommendations 1988 consist of the following documents:

- F.400 Message Handling System and Service Overview;
- F.401 Message Handling Services, Naming and Addressing for Public Message Handling Services;
- F.410 Message Handling Services, The Public Message Transfer Service;
- F.415 Message Handling Services, Intercommunication with Public Physical Delivery Services;
- F.420 Message Handling Services, The Public Interpersonal Messaging Service;
- F.421 Message Handling Services, Intercommunication between the IPM Service and the Telex Service;
- F.422 Message Handling Services, Intercommunication between the IPM Service and the Teletex Service.

2.2.2
The Functional Model

The principal changes to the 1984 MHS Functional Model in the 1988 version are the addition of the following two elements:

- Message Store;
- Access Unit.

2.2.2.1
Message Store

The Message Store (MS) is responsible for the storage and retrieval of messages on behalf of the UA. The MS is an optional element and, if available, is allocated between the UA and the corresponding MTA. Furthermore the MS enables the user to access his mailbox from a remote UA, when he is out of his office.

2.2.2.2
Access Unit

The Access Unit (AU) provides an interface to other non-MHS services such as other telematic services and physical delivery services.

The following figure shows the elements of the functional model of MHS 1988:

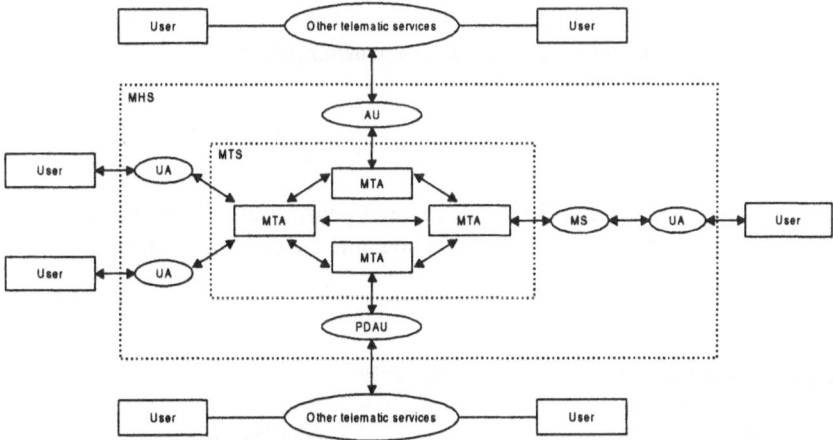

Fig. 2.2. Functional model of 1988 X.400 systems

2.2.3
Naming and Addressing

As mentioned earlier the naming and addressing conventions used in MHS 1984 comprise some problems and therefore had to be changed in the new version of X.400. In the MHS 1988 standard, an O/R name or address can either refer to a user person or process (represented in MHS by a UA and possibly a MS) or to a distribution list (DL). Three kinds of MHS name and address objects are defined:

- Directory names;
- O/R addresses;
- O/R names.

2.2.3.1
Directory Names

The directory name identifies who the sender or receiver of the message is. It is defined as a hierarchical list of relative distinguished names (RDNs) that together make up a unique identifier of the O/R user or DL. Taken together each directory name is unambiguous.

2.2.3.2
O/R Addresses

The O/R address identifies where (i.e. which domain and organisational unit) the sender or receiver is located and is a logical address. It specifies a MHS MD and within that domain an O/R user (or DL expansion point).

2.2.3.3
O/R Names

The O/R name defined in 1988 allows both naming and addressing information to be carried. It is a data structure that contains the directory name only, the O/R address only, or both.

2.2.4
New Features Introduced in MHS 1988

In MHS 1988 many additional features were added. These are briefly described below.

- Distribution Lists (DLs)
- Security
- Use of a Directory

- Messaging system types and Protocols

2.2.4.1
Distribution Lists (DLs)

Basically a DL is a function or role group name, as distinguished from the name of an individual. Several individuals acting on the same role group may need to get the same message. The members of the role group are determined by an individual called the DL owner, who is responsible for establishing and dynamically maintaining the group membership list. He may also restrict who is allowed to send messages to the DL, and may elect to serve as the focal point for collecting delivery and non-delivery reports, instead of letting the reports go back to the message originator.

2.2.4.2
Security

The MHS 1988 standard has an extensive list of security capabilities. These capabilities are provided in the nature of a toolkit, to be used by security authorities according to the security policies of the involved systems.

The security threats for a MHS are classified into four categories in the standard as follows:

- **access threats:** Invalid user access is obviously one of the prime security threats to the system. If an invalid user can be prevented from using the system, many of the other threats are substantially reduced.
- **inter-message threats:** Inter-message threats arise from unauthorised agents who are external to the message communication, and can manifest themselves by masquerading as a legitimate user and thus eavesdropping on legitimate messages, modifying messages, or replaying messages to cause confusion or extract more information from the indented recipient.
- **intra-message threats:** Intra-message threats are those arising from the actual message communication participants themselves, and can manifest themselves by denying involvement in the communication or by sending or receiving messages for which they have an inadequate security clearance.
- **data store threats:** MHS has many data stores within it that must be protected from threats, like modification of routing information, or copying a deferred delivery message while it is being held and "preplaying" it to cause confusion. MHS provides many capabilities to counter the above threats.

2.2.4.3
Use of a Directory

A Directory is a collection of open systems that provide directory services and is used by MHS for several different purposes:

- **user friendly naming:** The originator and recipient of a message can be identified by directory names, rather than by machine-orientated O/R addresses.
- **user capabilities:** The capabilities of a recipient (or originator UA), such as the content types and length of messages that it can accept or the encoded information types it can interpret, can be stored in the directory entry for each user.
- **DL expansion:** A group whose membership is stored in the directory can be used as a DL. The originator simply supplies the name of the list. At the DL expansion point the MTA can obtain the directory names (and the O/R addresses) of the individual users.
- **Authentication:** Before two MHS functional entities (e.g. two MTAs or UA and MTA) communicate with one another, each one has to identify himself. This can be done by using authentication capabilities of MHS based on information stored in the directory.

2.2.4.4
Messaging System Types and Protocols

With the addition of the MS as a new MHS service in 1988, seven distinct messaging system types can now be identified, corresponding to the seven different ways in which the UA, MS and MTA can be combined in one system:

- stand-alone UA;
- MTA together with zero or more UAs in one system;
- MTA together with a UA and zero or more UAs in one system;
- MTA together with a MS and zero or more UAs in one system;
- MTA together with both a UA and a MS and zero or more UAs in one system;
- stand-alone MS;
- UA together with a MS in one system.

In MHS 1984 only the P1, P2 and P3 protocols were available. The principal limitation of the P3 protocol is that it does not permit selective retrieval of messages awaiting delivery. Transfer priority is the only determinant, apart from arrival time, that is used by the MTA to decide which message to send next. This is obviously a problem for many users who want to have much finer control over the order in which they inspect newly arrived message.

In the 1988 series of recommendations a solution to this problem is provided by the MS and its associated Access Protocol P7. The messages are delivered to the MS in the normal way, but may then be interrogated by the UA. P7 allows a UA to list the messages in a MS and to fetch all or a part of them. It may also be used to help correlate delivery reports with submitted messages and to perform actions on behalf of the UA.

2.2.5
Downgrading X.400 (88) to X.400 (84)

Although almost all products available on the market are at least X.400 1988 compliant, MHS 1984 must be considered as well because a huge number of service providers only offer 1984 systems to their users. As a consequence of this MHS 1988 systems have to ensure that communication with the obsolete systems is possible without loss of information. It should be noted that as MHS 1988 is a strict superset of MHS 1984 the mapping between the two versions is only a one-way problem with one exception. In X.400 1988 size constraints have been defined for a number of MTS transfer protocol (P1) elements. Provided that a 1984 system observes these constraints, a correctly encoded Message Transfer System Application Protocol Data Unit (MTS-APDU) received from a 1984 system also conforms to 1988 MTS protocol (P1). Because of this a 1988 system does not have to take special actions when receiving a mail of a 1984 system. In Annex B of X.419 (1988) there are some downgrading rules but they are not sufficient for environments containing both 1984 and 1988 components. There are a few critical aspects when a message from a 1988 systems enters a 1984 system.

2.2.5.1
Service Irregularities

The use of redirection and distribution lists in the presence of 1988/84 domain boundaries may for instance lead to some service irregularities which are listed below:

- recipients may not be able to notice that they received a message because of DL expansion or redirection;
- when a message traverses a 1984 domain, the expansion history and the redirection history are lost. This may cause premature routing hop detection and result in redirection or expansion failure. Note that only a DL with a 1984 compatible O/R address may encounter this problem;
- 1984 MTAs will return notifications to the message originator rather than redirecting them back along the DL expansion path;
- 1984 systems may see new distinguished values for integer protocol elements which are unknown to them.

2.2.5.2
Avoiding Downgrading

Perhaps the most important consideration is to configure systems so as to minimise the need for downgrading. It is strongly recommended not to use 1984 systems to interconnect two or more 1988 systems.

In practise, many of the downgrading issues will be avoided. When a 1988 originator sends a mail to a 1984 recipient, 1988 specific features will not be used as they do not work! For DLs with 1984 and 1988 recipients, messages will tend to be "lowest common denominator".

2.2.5.3 Addressing

Because of the changes made in MHS 1988 regarding the O/R names and O/R addresses problems can arise in respect to addressing when 1988 specific features in O/R addresses are used. The X.419 approach will mean that addressing these features cannot be specified from 84 systems. Worse, a message originated from such a message cannot be transferred into a 1984 system. This is unacceptable. There are two approaches defined to solve this problem. The first one is a general purpose mechanism, which can be implemented by the gateway only. The second one is a special purpose mechanism that is optimised for a form of X.400 (88) address which is expected to be used frequently (Common Name). The second approach requires cooperation from all X.400 (88) UAs and MTAs which are involved in these interactions.

2.2.5.4 General Approach

The first approach is to use a domain defined attribute (DDA) "X.400-88". The DDA value is an "std-or encoding" of the address as defined in RFC 1327. This will allow source routing through an appropriate gateway. This solution is general and does not require the cooperation of the involved parties.

The std-or syntax cannot use IA5 characters in the printable string set (typically to handle teletex versions). To enable this, the std-or encoding is encapsulated into printable string using the mapping of section 3.4. of RFC 1327. Where the generated address is longer than 128 characters, up to three overflow DDAs are used:

- X.400-C1;
- X.400-C2;
- X.400-C3.

2.2.5.5
Common Name

Where a common name attribute is used, this is downgraded to the DDA "Common". For example:

X.400 (88):
> CN=Postmaster; O=A;ADMD=B;C=GB;

X.400 (84):
> DDCommon=Postmaster; O=A;ADMD=B;C=GB.

The downgrade will always happen correctly. However, it will not always be possible for the gateway to do the reverse mapping. Therefore all 1988 MTAs and UAs which wish to interact with 1984 systems through gateways following this specification will need to understand the equivalence of these two forms of address.

2.2.5.6
Message Transfer System

Annex B of X.419 is sufficient apart from addressing.

The discard of envelope fields is unfortunate. However, the criticality mechanism ensures that no information the originator specifies as critical is discarded. There is no sensible alternative. If mapping of a system which supports the MOTIS-86 trace extensions is performed, it is recommended that the internal trace of X.400 (88) is mapped on to these trace extensions, noting slight differences in syntax.

2.2.5.7
IPM Downgrading

The IPM service in X.400 (84) is usually provided by content type 2 whic differs from X.400 (88), where it is provided by content type 22. In many cases, it will be useful for a gateway to downgrade P2 from content type 22 to 2. This will clearly need to be made dependent on the destination, as it is quite possible to carry content type 22 over P1 (84). The decision to make this downgrade will have to be made on the basis of gateway configuration.

When a gateway downgrades from content type 22 to 2 the following should be done:

- strip any 1988 specific headings (language indication and partial message indication);
- downgrade all O/R addresses, as described above.

If a directory name is present, there is no method to preserve the semantics within a 1984 O/R address. However, it is possible to pass information across, so that the information in the distinguished name can be informally displayed to the end user. This is done by appending a text representation of the distinguished name to the free form name enclosed in round brackets. It is recommended that the "user friendly name" syntax is used to represent the distinguished name. For example:

(Gerhard Schmied, MHS Development, Infonova, AT)

The issue of body part downgrading is discussed below.

RFC 822 Considerations. A message represented as content type 22 may have originated from RFC 822. The downgrade for this type of message can be improved. This is discussed in RFC 1327.

Body Part Downgrading. The way how body part downgrading should be done is described in RFC 1496 where the following scenarios are considered:

- SMPT (MIME) -> X.400 (84);
- X.400 (84) -> SMTP (MIME).

The approach that is used in RFC 1496 is that the connection X.400 (84) <-> SMTP (MIME) never happens. This, of course, is an illusion, but it can be a very useful one. Instead of going the direct way all messages go one of the following paths:

- X.400 (84) -> X.400 (88) -> SMTP (MIME);
- SMTP (MIME) -> X.400 (88) -> X.400 (84).

when they are exchanged between a MIME and an X.400 (84) user. The approach at the interface between X.400 (88) and X.400 (84) is

- convert what can be converted;
- encapsulate what can be encapsulated;
- never drop a message.

Of course for X.400 (88) body parts that are already defined in X.400 (84), no downgrading should be done. In particular, multibody messages should remain multibody messages, IA5 messages including IA5 messages encoded as Extended Body Parts (EBPs), should remain IA5 messages, and G3Fax messages should remain G3Fax messages.

Conversion rules. Some body parts are defined by X.400 (88) as having both Basic and Extended forms. These are listed in Annex B of X.420.

For all of these, the transformation from the EBP to the Basic Body Part takes the form of putting the PARAMETERS and the DATA members together in a SEQUENCE.

This transformation should be applied by the gateway in order to allow (for example) X.400 (88) systems that use the Extended form of the IA5 body part to communicate with X.400 (84) systems.

Encapsulation format. For any body part that cannot be used directly by X.400 (84), the following IA5 body part is made:

- Content = IA5String
- First bytes of content: (the description is in USASCII, with C escape sequences used to represent the control characters):
- MIME version: >version \r\n
- Content type : <the proper MIME content type>\r\n
- Content-transfer-encoding: <quoted printable or base64>\r\n
- <Possibly other Content headings here, terminated by\r\n>\r\n
- <Here follows the bytes of the content encoded in the proper encoding>

All implementations must place the "MIME-version:" header first in the body part. Headers that are placed by RFC 1327 and RFC 1496 into other parts of the message must not be placed in the MIME body part.

This includes RFC-822 headings carried as heading extensions, which must be placed in a new IA5 body part starting with the string "RFC-822-HEADERS", as specified in RFC 1327, Appendix G.

Other header extensions are still handled as described in chapter 5 of RFC 1327: They are dropped.

Since all X.400 (88) body parts can be represented in MIME by using the X.400-bp-MIME content-type, this conversion will never fail. In the reverse direction, any IA5 body part that starts with the token "MIME-version:" will be subject to conversions according to RFC 1496 before including the body part into an X.400 (88) message.

2.3
The X.400 Standard 1992

The 1992 X.400 Series of recommendations (actually approved in 1993) are primarily a re-issue of the 1988 series of recommendations with the rectification of the problems reported on the 1988 standards. The problems included both editorial and minor technical changes, minor in the sense that no fundamental changes to services or protocols are made through problem reports. There are however a number of important minor changes:

- a file transfer body part has been added to IPMS;
- a new voice body part has been added to IPMS;
- auto-submitted EoS;
- MS register EoS.

The organisation of the standard has changed as well. In the 1992 version of the standard the system and the service description are no longer together i one series of recommendations. The F.400 series of recommendations deals with the services specifications and the X.400 series of recommendations deals with specifying the system . Below is a listing of the F.400 series of recommendations and the X.400 series of recommendations of MHS 1992:

2.3.1
F.400 Series of Recommendations 1992

- F.400 Message Handling System and Service Overview
- F.401 Message Handling Services, Naming and Addressing for Public Message Handling Services
- F.410 Message Handling Services, the Public Message Transfer Service
- F.415 Message Handling Services, Intercommunication with Public Physical Delivery Services
- F.420 Message Handling Services, the Public Interpersonal Messaging Service
- F.421 Message Handling Services, Intercommunication between the IPM Service and the Telex Service
- F.422 Message Handling Services, Intercommunication between the IPM Service and the Teletex Service
- F.423 Message Handling Services, Intercommunication between the IPM Service and Telefax Service
- F.435 Message Handling Services, EDI Messaging Service
- F.440 Message Handling Services, Voice Messaging Service

2.3.2
X.400 Series of Recommendations of MHS 1992

- X.402 Message Handling Systems: Overall architecture
- X.403 Message Handling Systems: Conformance testing
- X.408 Message Handling Systems: Encoded information type conversion rules
- X.411 Message Handling Systems: Message Transfer System: Abstract service definition and procedures
- X.413 Message Handling Systems: Message Store: Abstract service definition
- X.419 Message Handling Systems: Protocol specifications
- X.420 Message Handling Systems: Interpersonal messaging system
- X.435 Message Handling Systems: EDI messaging system

* X.440 Message Handling Systems: Voice messaging system

2.4
The X.400 Standard 1996

The enhancements introduced in this new version of the X.400 standards refer to a great degree to additional features of the message store and the interpersonal messaging. Many of the new EoS deal with auto-actions that can be performed by users.

There were no changes made to the organisation and naming of the standards so that they are still the same as in MHS 1992.

2.5
Proposed Additions to the X.400 Functionality for Multimedia Messaging

The X.400 series of recommendations (1996) supports multimedia messages (text, voice, video,...). However the real operation in a "multimedia/hypermedia" fashion is not yet standardised neither in ITU-T/ISO nor in the relevant RFCs. Because of the fact that multimedia data are generally very large it is very useful not always to transmit the multimedia content as part of the message, but to provide a hyperlink functionality so that multimedia contents are only transmitted on demand. This would significantly reduce network traffic especially when large amounts of multimedia contents are sent to a huge number of users via distribution lists.

These features could be implemented in a way where the text body part of the message is scanned for the strings "http://" or "ftp://". If one of these strings occurs, the string until the next blank is highlighted. If the users click on that highlighted string the multimedia data are transmitted to the users' desktop.

2.6
Physical Access to the X.400 Service

The following section gives a brief overview of important network access types for X.400 services.

2.6.1
Public Switched Telephone Network (PSTN)

Via modems data transmission rates up to 34400 kbps can be achieved. Maximal bit rates, however, may be reduced due to international connectivity reasons to 9.6 kbps or even lower.

2.6.2
Integrated Services Digital Network (ISDN)

A separate WP1 deliverable "Service Description of the X.400 Service for Euro-ISDN" will specify in detail how the X.400 service can be supported in the most optimised way by Euro-ISDN.

A Basic Rate Interface (BRI) is two 64K bearer ("B") channels and a single delta ("D") channel. The B channels are used for voice or data, and the D channel is used for signalling and/or X.25 packet networking. The Packet Mode Bearer Service (PMBS) over B and D channels, an ITU standard, offer access for user equipment based on X.25 protocols.

Equipment known as a Terminal Adapter (TA) can be used to adapt these channels to existing terminal equipment standards such as RS-232/V.34 and V.35/V.36. This equipment is typically packaged in a similar fashion to modems, either as stand-alone units or as interface cards that plug into a computer or various kinds of communications equipment (such as routers or PBXs).

The following data transmission rates are possible (in Europe):

Basic rate	2B+1D	2x64kbps (B-channels)	16 kbps (D-channel);
Primary rate	30B+1D	30x64 kbps	64 kbps.

B-channels are circuit switched, D-channels are packet switched. The D-channel is used for exchanging control information.

2.6.2.1
Relevant Standards on ISDN

There are numerous ITU-T (formerly CCITT) standards on ISDN.

Q.921:	"ISDN User-Network Interface Data Link Layer Specifications", 1988. The D-channel protocol. Found in Blue book Fascicle VI.10
Q.931:	"ISDN User-Network Interface Layer 3 Specification for Call Control", 1988. The messages that are sent over the D channel to set up calls, disconnect calls, etc. Found in Blue book Fascicle VI.11
Q.930:	General Overview
Q.932:	Generic procedures for the control of ISDN supplementary services
Q.933:	Frame Mode Call Control
G.711:	Pulse Code Modulation (PCM) of Voice Frequencies
G.722:	7-kHz Audio Coding Within 64 kbit/s
G.728:	Coding of Speech at 16 kbit/s using Low-Delay Code Excited Linear Prediction (LD-CELP)

H.320:	Narrow-band Visual Telephone Systems and Terminal Equipment
H.221:	Frame Structure for a 64 to 1920 kbit/s Channel in Audiovisual Teleservices
H.230:	Frame Synchronous Control and Indication Signals for Audiovisual Systems
H.242:	System for Establishing Communications between Audiovisual Terminals using Digital Channels up to 2 Mbit/s
H.261:	Video Codec for Audiovisual Services at p x 64 kbits/s
H.243:	Basic MCU Procedures for Establishing Communications between Three or More Audiovisual
I.2xy	ISDN Frame Mode Bearer Services, 1990
I.310	ISDN - Network Functional Principles
I.320	ISDN protocol reference model
I.324	ISDN Network Architecture
I.325	Reference configs for ISDN connection types
I.330	ISDN numbering and addressing principles
I.331	Numbering plan for ISDN (and several more in I.33x relating to numbering and addressing and routing)
I.340	ISDN connection types
I.350/351/352	Refer to performance objectives
I.410-412	Refer to user-network interfaces as I.420 and 421
I.430/430	Layer 1 specs
I.440/441	Layer 2 specs (Q.921)
I.450-452	Layer 3 specs (Q.931)
I.450	General Overview
I.451	Basic ISDN call control
I.452:	Extensions
I.460-465	Multiplexing and rate adaption
I.470	Relationship of terminal functions to ISDN
V.110:	"Support of DTE's with V Series Type Interfaces by an ISDN" Terminal rate adaption by bit stuffing. C.f. V120.
V.120:	"Support by an ISDN of Data Terminal Equipment with V Series Type Interfaces with Provision for Statistical Multiplexing" 1990 (This has been amended since the blue book)

2.6.3
Asynchronous Transfer Mode (ATM)

ATM (Asynchronous Transfer Mode) is a switching/transmission technique where data is transmitted in small, fixed sized cells (5 byte header, 48 byte payload). The cells lend themselves both to the time-division-multiplexing characteristics of the transmission media, and the packet switching characteristics desired of data networks.

At each switching node, the ATM header identifies a "virtual path" or "virtual circuit" that the cell contains data for, enabling the switch to forward the cell to the correct next-hop trunk. The "virtual path" is set up through the involved switches when two endpoints wish to communicate. This type of switching can be implemented in hardware, almost essential when trunk speed range from 45Mb/s to 1Gb/s.

New standards on ATM (PINNI 1.0) enable multivendor ATM networks. They define protocols for combination of different ATM switches.

One use of ATM is to serve as the core technology for a new set of ISDN offerings known as Broadband ISDN (B-ISDN). It is recommended to have further study on this technique within ISDN related parts of the project.

2.6.3.1
LAN Emulation (LANE)

A way for legacy LAN MAC-layer protocols like Ethernet and token ring, and all higher-layer protocols and applications, to access a network transparently across an ATM network. LAN emulation retains all Ethernet and token ring drivers and adapters; no modifications need to be made to Ethernet or token ring end station.

2.6.3.2
Multiple Protocol over ATM (MPOA)

In essence, MPOA expands on schemes like LAN emulation (LANE) from the ATM forum, as well as classical IP over ATM, next-hop routing protocol (NHRP), and c (MARS) from the IETF (Internet Engineering Task Force). Each of these approaches solves a piece of the ATM internetworking problem; what makes MPOA different is its ability to integrate these solutions into a unified whole. What's more, it adds a new concept: virtual routers.

The MPOA architecture comprises the following components:

- **Edge devices:** Sometimes referred to as multilayer switches, edge devices are intelligent switches that use either the destination's network-layer address or its MAC-layer address to forward packets between legacy LAN segments and ATM interfaces.
- **ATM-attached hosts:** These are ATM adapter cards that implement the MPOA protocol as part of their drivers. They let ATM-attached hosts communicate efficiently with one another or with legacy LANs connected by an edge device.
- **Route servers:** Not physical devices as such, route servers are a collection of functions that make it possible for network-layer subnets to be mapped onto ATM. Route servers can be implemented as stand-alone products, or they can consist of software added to existing routers or switches.

2.6.4
Packet Switched Data Networks (PSDN) X.25 and Frame Relay

Public and private X.25 networks are spread out all over the world. X.25 networks are interconnected by means of ITU-specified gateways (X.75 protocol), resulting in a worldwide X.25 network. Numbering is according to ITU Rec. X.121, routing according to X.110. A disadvantage of X.25 network is available speed, which is at its best 64 kbps. Only pure data transmission is enabled; it is not possible to transmit voice or video information.

The second generation packet switching technique, Frame Relay, provides a maximal bitrate of 2 Mbps. Frame Relay is often considered to be a less expensive but flexible alternative to ATM.

Whereas it is not possible to split one physical ISDN-B channel out in several logical channels, this could be done with frame relay technique. Thus, several hundred logical connection could be run over one physical T1 connection. However, ISDN could be used to connect the user to the Frame Relay net.

Interworking between X.25, Frame Relay and ATM is under further study worldwide. American companies such as AT&T or MCI, and also European companies such as BT or Unisource are providing Frame Relay based systems.

2.6.5
Global System for Mobile Communication (GSM)

GSM was primarily developed as a European standard for mobile telecommunication. Meanwhile networks are set up all over the world, based on the GSM standards.

An interface between GSM and ISDN, the so-called ISUP ("ISDN User Part") for basic and enhanced services, has recently been specified in ITU-T Recommendations Q.761 - Q.766.

3 Internet Mail

The history of Internet mail goes back to the early 70s and mailing has been one of the first "hot applications" on the Internet. Similar to the exchange of information in the normal way, in the Internet mail environment information is exchanged between persons. The most important reason why the importance of email has improved in such a significant way is the fact that it is very much faster and cheaper than the conventional post. But nevertheless both ways share many properties.

In the late 70s and early 80s the first standards for the format of Internet mail messages (RFC 822) and the way they can be exchanged (RFC 821) arose. Mailing relies on the store-and-forward principle. This means that, rather than to establish a direct connection between the originator's and the recipient's host, the message may be relayed by several intermediate hosts. At each host, the message is stored in order to find out whether it is addressed to a recipient connected to this host. In the case where the recipient is connected to the host the message is delivered to the recipient's mailbox. Otherwise it is relayed to the next host on the way to the recipient's host. In order to make the right routing decision the hosts use routing tables.

The remainder of this chapter deals with the most important Internet Request for Commence (RFCs) for Internet mail.

3.1 RFC 822: Standard for the Format of ARPA Internet Text Messages

RFC 822 is one of the most important RFCs concerning Internet mail because it specifies the format each message must possess in the Internet mail environment. It was written in 1982 and is the successor of RFC 733 describing the Standard for the ARPA Network Text Messages. As can be seen in the title of both RFCs in the beginning the standards only consider text messages.

Messages are viewed as having an envelope and contents. The envelope contains whatever information is needed to accomplish transmission and delivery. The contents compose the object to be delivered to the recipient. RFC 822 applies only to the format and some of the semantics of message contents. It contains no specification of the information in the envelope. However, some message systems may use information from the contents to create the envelope. Some message systems may store messages in formats that differ from the one specified in RFC 822 which is intended strictly as a definition of what message content format is to be passed BETWEEN hosts.

Messages can be made complex and rich with formally structured components of information or can be kept small and simple, with a minimum of such information. RFC 822 also simplifies the interpretation of differing visual formats in messages, only the visual aspect of a message is affected and not the interpretation of information within it.

3.1.1
Lexical Analysis of Messages

3.1.1.1
General Description

As already mentioned above, a message consists of header fields and, optionally, a body. The body is simply a sequence of lines containing ASCII characters. It is separated from the headers by null lines [i.e. a line with nothing preceding the Carriage Return Line Feed (CRLF)].

3.1.1.2
Long Header Fields

Each header field can be viewed as a single, logical line of ASCII characters, comprising a field-name and a field-body. For convenience, the field-body portion of this conceptual entity can be split into a multiple-line representation. This action is called "folding". The general rule is that wherever there may be linear-white-space (not simply LWSP-chars), a CRLF immediately followed by AT LEAST one LWSP-char may instead be inserted. Thus, the single line

        To: "Joe & J. Harvey" <ddd @Org>, JJV @ BBN

can be represented as

        To: "Joe & J. Harvey" <ddd @ Org>,
        JJV@BBN

and

        To: "Joe & J. Harvey"
        <ddd@ Org>, JJV
        @BBN

and

To: "Joe &
     J. Harvey" <ddd @ Org>, JJV @ BBN

The process of moving from this folded multiple-line representation of a header field to its single-line representation is called "unfolding". Unfolding is accomplished by regarding CRLF immediately followed by a LWSP-char as equivalent to the LWSP-char.

Note that although folding is allowed wherever linear-white-space is permitted, it is recommended that structured fields, such as those containing addresses, limit folding to higher-level syntactic breaks. For address fields, it is recommended that such folding occur between addresses, after the separating comma.

3.1.1.3
Structure of Header Fields

Once a field has been unfolded, it may be viewed as being composed of a field-name followed by a colon (":"), followed by a field-body, and terminated by a CRLF. The field-name must be composed of printable ASCII characters (i.e. characters that have values between 33 and 126. decimal, except colon). The field-body may be composed of any ASCII characters, except CR or LF. (While CR and/or LF may be present in the actual text, they are removed by the action of unfolding the field.)

Certain field-bodies of headers may be interpreted according to an internal syntax that some systems may wish to parse.

These fields are called "structured fields". Examples include fields containing dates and addresses. Other fields, such as "Subject" and "Comments", are regarded simply as strings of text. Any field which has a field-body that is defined as other than simply <text> is to be treated as a structured field.

Field-names, unstructured field-bodies and structured field-bodies are each scanned by their own, independent "lexical" analysers.

3.1.1.4
Unstructured Field-Bodies

For some fields, such as "Subject" and "Comments", no structuring is assumed, and they are treated simply as <text>s, as in the message body. Because of the fact that rules of folding apply to these fields as well, such field-bodies which occupy several lines must therefore have the second and successive lines indented by at least one LWSP-char.

3.1.1.5
Structured Field-Bodies

To aid in the creation and reading of structured fields, the free insertion of linear-white-space (which permits folding by inclusion of CRLFs) is allowed between lexical tokens. Rather than obscuring the syntax specifications for these structured fields with explicit syntax for this linear-white-space, the existence of another "lexical" analyser is assumed. This analyser does not apply for unstructured field-bodies that are simply strings of text, as described above. The analyser provides an interpretation of the unfolded text composing the body of the field as a sequence of lexical symbols. These symbols are

- individual special characters
- quoted strings
- domain literals
- comments
- atoms

In contrast to the first four of these symbols which are self-delimiting atoms are delimited by the self-delimiting symbols and by linear-white-space. For the purposes of regenerating sequences of atoms and quoted strings, exactly one SPACE is assumed to exist, and should be used, between them.
So, for example, the folded body of an address field

> ":sysmail"@ Some-Group. Some-Org,
> Muhammed.(I am the greatest) Ali @(the)Vegas.WBA

is analysed into the following lexical symbols and types:

:sysmail	quoted string
@	special
Some-Group	atom
.	special
Some-Org	atom
,	special
Muhammed	atom
.	special
(I am the greatest)	comment
Ali	atom
@	atom
(the)	comment
Vegas	atom
.	special
WBA	atom

The canonical representations for the data in these addresses are the following strings:

">:sysmail"@Some-Group.Some-Org

and

Muhammed.Ali@Vegas.WBA

Note: For purposes of display, and when passing such structured information to other systems, such as mail protocol services, there must be NO linear-white-space between <word>s that are separated by period (".") or at-sign ("@") and exactly one SPACE between all other <word>s. Also, headers should be in a folded form.

3.1.1.6
Header Field Definitions

These rules show a field meta-syntax, without regard for the particular type or internal syntax. Their purpose is to permit detection of fields and additionally they present an image of each field as fitting on one line to higher-level parsers.

```
field = field-name ":" [field-body] CRLF
field-name = 1*<any CHAR, excluding CTLs, SPACE, and ":">
field-body = field-body-contents [CRLF LWSP-char field-body]
field-body-contents = <the ASCII characters making up the field-body,
                        as defined in the following sections, and
                        consisting of combinations of atoms,
                        quoted strings, and special tokens, or else
                        consisting of texts>.
```

3.1.1.7
Lexical Tokens

The following rules are used to define an underlying lexical analyser, which feeds tokens to higher-level parsers.

		(Octal, Decimal)
CHAR	= <any ASCII character>	(0-177, 0.-127)
ALPHA	= <any ASCII alphabetic character>	(101-132, 65.- 90)
		(141-172, 97.-122)
DIGIT	= <any ASCII decimal digit>	(60- 71, 48.- 57)
CTL	= <any ASCII control character and DEL>	(0- 37, 0.- 31)
		(177, 127)
CR	= <ASCII CR, carriage return>	(15, 13)
LF	= <ASCII LF, linefeed>	(12, 10)

```
SPACE = <ASCII SP, space>                    (40, 32)
HTAB = <ASCII HT, horizontal-tab>            (11, 9)
<"> = <ASCII quote mark>                      (42, 34)
CRLF = CR LF
LWSP-char = SPACE / HTAB                      semantics = SPACE
linear-white-space = 1*([CRLF] LWSP-char)    semantics = SPACE
                                              CRLF => folding
specials = "(" / ")" / "<" / ">" / "@"        Must be in quoted
           / "," / ";" / ":" / "\" / <"> ;    string, to use within a word
           / "." / "[" / "]"
delimiters = specials / linear-white-space /
             comment
text = <any CHAR, including bare              => atoms, specials,
       CR & bare LF, but not                  comments and
       including CRLF>                        quoted strings are not
                                              recognised.
atom = 1*<any CHAR except specials, SPACE and CTLs>
quoted string = <"> *(qtext/quoted pair) <">  Regular  qtext  or  quoted
                                              chars
qtext = <any CHAR excepting <">,              => may be folded
        "\" & CR, and including
        linear-white-space>
domain literal = "[" *(dtext / quoted-pair) "]"
dtext = <any CHAR excluding "[","]", "\" & CR, => may be folded
        & including linear-white-space>
comment = "(" *(ctext / quoted pair / comment) ")"
ctext = <any CHAR excluding "(", ")", "\" & CR, => may be folded
        ")", "\" & CR, & including
        linear-white-space>
quoted-pair = "\" CHAR                        may quote any char
phrase = 1*word                               Sequence of words
word = atom / quoted string
```

A detailed description and clarification of some of these lexical tokens can be found in RFC 822.

3.1.2
Message Specification

3.1.2.1
Syntax

Due to an artifact of the notational conventions, the syntax indicates that, when present, some fields must be in a particular order. Header fields are not required to occur in any particular order, except that the message body must occur AFTER the headers. It is recommended that, if present, headers be sent in the order "Return-Path", "Received", "Date", "From", "Subject", "Sender", "To", "cc", etc.

RFC 822 permits multiple occurrences of most fields. Except as noted, their interpretation is not specified here, and their use is discouraged.

The following syntax for the bodies of various fields should be thought of as describing each field-body as a single long string (or line). As indicated above such long strings can be represented on more than one line in the actual transmitted message.

message = fields *(CRLF *text)	Everything after first null line is message body
fields = dates	Creation time,
source	author id & one
1*destination	address required
*optional-field	others optional
source = [trace]	net traversals
originator	original mail
[resent]	forwarded
trace = return	path to sender
1*received	receipt tags
return = "Return-path" ":" route-addr	return address
received = "Received" ":"	one per relay
["from" domain]	sending host
["by" domain]	receiving host
["via" atom]	physical path
*("with" atom)	link/mail protocol
["id" msg-id]	receiver msg id
["for" addr-spec]	initial form
";" date-time	time received
originator = authentic	authenticated addr

```
                    ["Reply-To" ":" 1#address])

authentic =        "From" ":" mailbox              Single author
           / ( "Sender" ":" mailbox               Actual submittor
               "From" ":" 1#mailbox)              Multiple   authors   or   no
                                                  senders

resent = resent-authentic
        ["Resent-Reply-To"            ":"        1#address])

resent-authentic =
           =       "Resent-From"     ":"         mailbox
           /   ( "Resent-Sender"  ":"            mailbox
               "Resent-From"     ":"             1#mailbox)

dates = orig-date               Original
        [resent-date]                            Forwarded

orig-date = "Date" ":" date-time

resent-date = "Resent-Date" ":" date-time

destination =     "To"            ":" 1#address    Primary
           /    "Resent-To"  ":"                  1#address
           /    "cc"            ":" 1#address      Secondary
           /    "Resent-cc"  ":" 1#address
           /    "bcc"          ":" #address        Blind carbon
           /    "Resent-bcc" ":" #address

optional-field =
           /    "Message-ID"          ":"         msg-id
           /    "Resent-Message-ID" ":"           msg-id
           /    "In-Reply-To"         ":"         *(phrase / msg-id)
           /    "References"          ":"         *(phrase / msg-id)
           /    "Keywords"            ":"         #phrase
           /    "Subject"             ":"         *text
           /    "Comments"            ":"         *text
           /    "Encrypted"           ":"         1#2word
           /    extension-field                  To be defined
           /    user-defined-field               May be pre-empted

msg-id = "  <  " addr-spec  ">"                   Unique message id

extension-field =
           <Any field which is defined in a document published as a formal
           extension to RFC 822; none will have names beginning with the
           string "X-">
```

user-defined-field =
> <Any field which has not been defined in or published as an extension to RFC 822; names for such fields must be unique and may be pre-empted by published extensions>

3.1.2.2
Forwarding

Some systems permit mail recipients to forward a message, retaining the original headers, by adding some new fields. RFC 822 supports such a service, through the "Resent-" prefix to field-names.

Whenever the string "Resent-" begins a field-name, the field has the same semantics as a field whose name does not have the prefix. However, the message is assumed to have been forwarded by an original recipient who attached the "Resent-" field. This new field is treated as being more recent than the equivalent, original field. For example, the "Resent-From" indicates the person that forwarded the message, whereas the "From" field indicates the original author.

Use of such precedence information depends upon participants' communication needs. In general, the "Resent-" fields should be treated as containing a set of information that is independent of the set of original fields. Information for one set should not automatically be taken from the other. The interpretation of multiple "Resent-" fields, of the same type, is undefined.

3.1.2.3
Trace Fields

Trace information is used to provide an audit trail of message handling. In addition, it indicates a route back to the sender of the message.

The list of known "via" and "with" values are registered with the Network Information Center, SRI International, Menlo Park, California.

3.1.2.4
Return-Path

This field is added by the final transport system that delivers the message to its recipient. The field is intended to contain definitive information about the address and route back to the message's originator.

The "Reply-To" field is added by the originator and serves to direct replies, whereas the "Return-Path" field is used to identify a path back to the originator.

While the syntax indicates that a route specification is optional, every attempt should be made to provide that information in this field.

3.1.2.5
Received

A copy of this field is added by each transport service that relays the message. The information in the field can be quite useful for tracing transport problems.

The names of the sending and receiving hosts and time-of-receipt may be specified. The "via" parameter may be used to indicate over which physical mechanism the message was sent, such as Arpanet or Phonenet, and the "with" parameter may be used to indicate the mail-, or connection-, level protocol that was used, such as the SMTP mail protocol, or X.25 transport protocol. Several "with" parameters may be included to fully specify the set of protocols that were used.

Some transport services queue mail; the internal message identifier that is assigned to the message may be noted, using the "id" parameter. When the sending host uses a destination address specification that the receiving host reinterprets, by expansion or transformation, the receiving host may wish to record the original specification, using the "for" parameter. For example, when a copy of mail is sent to the member of a distribution list, this parameter may be used to record the original address that was used to specify the list.

3.1.2.6
Originator Fields

The standard allows only a subset of the combinations possible with the From, Sender, Reply-To, Resent-From, Resent-Sender, and Resent-Reply-To fields. The limitation is intentional.

3.1.2.7
From / Resent-From

This field contains the identity of the person(s) who wished this message to be sent. The message-creation process should default this field to be a single, authenticated machine address, indicating the AGENT (person, system or process) entering the message. If this is not done, the "Sender" field must be present. If the "From" field IS defaulted this way, the "Sender" field is optional and is redundant with the "From" field. In all cases, addresses in the "From" field must be machine-usable (addr-specs) and may not contain named lists (groups).

3.1.2.8
Sender / Resent-Sender

This field contains the authenticated identity of the AGENT (person, system or process) that sends the message. It is intended for use when the sender is not the author of the message, or to indicate who among a group of authors actually sent the message. If the contents of the "Sender" field would be completely redundant with the "From" field, then the "Sender" field need not be present and its use is discouraged (though still legal). In particular, the "Sender" field must be present if it is not the same as the "From" Field.

The Sender mailbox specification includes a word sequence which must correspond to a specific agent (i.e. a human user or a computer programme) rather than a standard address. This indicates the expectation that the field will identify the single AGENT (person, system, or process) responsible for sending the mail and not simply include the name of a mailbox from which the mail was sent. For example in the case of a shared login name, the name, by itself, would not be adequate. The local-part address unit, which refers to this agent, is expected to be a computer system term, and not (for example) a generalised person reference which can be used outside the network text message context.

Since the critical function served by the "Sender" field is identification of the agent responsible for sending mail and since computer programs cannot be held accountable for their behaviour, it is strongly recommended that when a computer programme generates a message, the HUMAN who is responsible for that programme be referenced as part of the "Sender" field mailbox specification.

3.1.2.9
Reply-To / Resent-Reply-To

This field provides a general mechanism for indicating any mailbox(es) to which responses are to be sent. Three typical uses for this feature can be distinguished. In the first case, the author(s) may not have regular machine-based mailboxes and therefore wish(es) to indicate an alternate machine address. In the second case, an author may wish additional persons to be made aware of, or responsible for, replies. A somewhat different use may be of some help to "text message teleconferencing" groups equipped with automatic distribution services: include the address of that service in the "Reply-To" field of all messages submitted to the teleconference; then participants can "Reply" to conference submissions to guarantee the correct distribution of any submission of their own.

3.1.2.10
Automatic Use of From / Sender / Reply-To

For systems which automatically generate address lists for replies to messages, the following recommendations are made:

- The "Sender" field mailbox should be sent notices of any problems in transport or delivery of the original messages. If there is no "Sender" field, then the "From" field mailbox should be used.
- The "Sender" field mailbox should NEVER be used automatically in a recipient's reply message.

- If the "Reply-To" field exists, then the reply should go to the addresses indicated in that field and not to the address(es) indicated in the "From" field.
- If there is a "From" field, but no "Reply-To" field, the reply should be sent to the address(es) indicated in the "From" field.

Sometimes, a recipient may actually wish to communicate with the person who initiated the message transfer. In such cases, it is reasonable to use the "Sender" address.

This recommendation is intended only for automated use of originator-fields and is not intended to suggest that replies may not also be sent to other recipients of messages. It is up to the respective mail-handling programmes to decide what additional facilities will be provided.

3.1.2.11
Receiver Fields

To / Resent-To. This field contains the identity of the primary recipients of the message.

Cc / Resent-Cc. This field contains the identity of the secondary (informational) recipients of the message.

Bcc / Resent-Bcc. This field contains the identity of additional recipients of the message. The contents of this field are not included in copies of the message sent to the primary and secondary recipients. Some systems may choose to include the text of the "Bcc" field only in the author(s)'s copy, while others may also include it in the text sent to all those indicated in the "Bcc" list.

3.1.2.12
Reference Fields

Message-ID / Resent-Message-ID. This field contains a unique identifier (the local-part address unit) which refers to THIS version of THIS message. The uniqueness of the message identifier is guaranteed by the host which generates it. This identifier is intended to be machine-readable and not necessarily meaningful to humans. A message identifier pertains to exactly one instantiation of a particular message; subsequent revisions to the message should each receive new message identifiers.

In-Reply-To. The contents of this field identify previous correspondence which this message answers. Note that if message identifiers are used in this field, they must use the msg-id specification format.

References. The contents of this field identify other correspondence which this message references. Note that if message identifiers are used, they must use the msg-id specification format.

Keywords. This field contains keywords or phrases, separated by commas.

3.1.2.13
Other Fields

Subject. This is intended to provide a summary, or indicate the nature, of the message.

Comments. Permits adding text comments onto the message without disturbing the contents of the message's body.

Encrypted. Sometimes, data encryption is used to increase the privacy of message contents. If the body of a message has been encrypted, to keep its contents private, the "Encrypted" field can be used to note the fact and to indicate the nature of the encryption. The first <word> parameter indicates the software used to encrypt the body, and the second, optional <word> is intended to aid the recipient in selecting the proper decryption key. This code word may be viewed as an index to a table of keys held by the recipient.

Note: Unfortunately, headers must contain envelope, as well as contents, information. Consequently, it is necessary that they remain unencrypted so that mail transport services may access them. Since names, addresses, and "Subject" field contents may contain sensitive information, this requirement limits total message privacy.

Extension-Field. A limited number of common fields have been defined in RFC 822. As network mail requirements dictate, additional fields may be standardised. To provide user-defined fields with a measure of safety, in name selection, such extension-fields will never have names that begin with the string "X-".

User-Defined-Field. Individual users of network mail are free to define and use additional header fields. Such fields must have names which are not already used in RFC 822 or in any definitions of extension-fields, and the overall syntax of these user-defined-fields must conform to rules specified in RFC 822 for delimiting and folding fields. Due to the extension-field publishing process, the name of a user-defined-field may be pre-empted. It should be noted that the prefatory string "X-" will never be used in the names of Extension-fields. This provides user-defined fields with a protected set of names.

3.1.3
Date and Time Specification

3.1.3.1
Syntax

date-time	=	[day ","] date time	dd mm yy hh:mm:ss zzz
day	=	"Mon" / "Tue" / "Wed" / "Thu" / "Fri" / "Sat" / "Sun"	
date	=	1*2DIGIT month 2DIGIT	day month year e.g. 20 Jun 82
month	=	"Jan" / "Feb" / "Mar" / "Apr" / "May" / "Jun" / "Jul" / "Aug" / "Sep" / "Oct" / "Nov" / "Dec"	
time	=	hour zone	ANSI and Military
hour	=	2DIGIT ":" 2DIGIT [":" 2DIGIT]	00:00:00 - 23:59:59
zone	=	"UT" / "GMT"	Universal Time, North American : UT
	/	"EST" / "EDT"	Eastern: - 5/ - 4
	/	"CST" / "CDT"	Central: - 6/ - 5
	/	"MST" / "MDT"	Mountain: - 7/ - 6
	/	"PST" / "PDT"	Pacific: - 8/ - 7
	/	1ALPHA	Military: Z = UT; A:-1; (J not used) M:-12; N:+1; Y:+12
	/	(("+" / "-") 4DIGIT)	Local differential hours+min.(HHMM)

3.1.3.2
Semantics

If included, day-of-week must be the day implied by the date specification.

Time zone may be indicated in several ways. "UT" is Universal Time (formerly called "Greenwich Mean Time"); "GMT" is permitted as a reference to Universal Time. The military standard uses a single character for each zone. "Z" is Universal Time. "A" indicates one hour earlier, and "M" indicates 12 hours earlier; "N" is one hour later, and "Y" is 12 hours later. The letter "J" is not used. The other remaining two forms are taken from ANSI standard X3.51-1975. One allows explicit indication of the amount of offset from UT; the other uses common 3-character strings for indicating time zones in North America.

3.1.4
Address Specification

3.1.4.1
Syntax

address	=	mailbox / group	one addressee / named list
group	=	phrase ":" [#mailbox] ";"	
mailbox	=	addr-spec	simple address
	/	phrase route-addr	name & addr-spec
route-addr	=	"<" [route] addr-spec ">"	
route	=	1#("@" domain) ":"	path-relative
addr-spec	=	local-part "@" domain	global address
local-part	=	word *("." word)	uninterpreted case-preserved
domain	=	sub-domain *("." sub-domain)	
sub-domain	=	domain-ref / domain-literal	
domain-ref	=	atom	symbolic reference

3.1.4.2
Semantics

A mailbox receives mail. It is a conceptual entity which does not necessarily pertain to file storage. For example, some sites may choose to print mail on their line printer and deliver the output to the addressee's desk.

A mailbox specification comprises a person, system or process name reference, a domain-dependent string, and a name-domain reference. The name reference is optional and is usually used to indicate the human name of a recipient. The name-domain reference specifies a sequence of sub-domains. The domain-dependent string is uninterpreted, except by the final sub-domain; the rest of the mail service merely transmits it as a literal string.

3.1.4.3
Domains

A name-domain is a set of registered (mail) names. A name-domain specification resolves to a subordinate name-domain specification or to a terminal domain-dependent string.

Hence, domain specification is extensible, permitting any number of registration levels.

Name-domains model a global, logical, hierarchical addressing scheme. The model is logical, in that an address specification is related to name registration and is not necessarily tied to transmission path. The model's hierarchy is a directed graph, called an in-tree, such that there is a single path from the root of the tree to any node in the hierarchy. If more than one path actually exists, they are considered to be different addresses.

The root node is common to all addresses; consequently, it is not referenced. Its children constitute "top-level" name-domains. Usually, a service has access to its own full domain specification and to the names of all top-level name-domains.

The "top" of the domain addressing hierarchy -- a child of the root -- is indicated by the right-most field, in a domain specification. Its child is specified to the left, its child to the left, and so on.

Some groups provide formal registration services; these constitute name-domains that are independent logically of specific machines. In addition, networks and machines implicitly compose name-domains, since their membership usually is registered in name tables.

In the case of formal registration, an organisation implements a (distributed) data base which provides an address-to-route mapping service for addresses of the form

person@registry.organisation

Note that "organisation" is a logical entity, separate from any particular communication network.

A mechanism for accessing "organisation" is universally available. That mechanism, in turn, seeks an instantiation of the registry; its location is not indicated in the address specification. It is assumed that the system which operates under the name "organisation" knows how to find a subordinate registry. The registry will then use the "person" string to determine where to send the mail specification.

The latter, network-oriented case permits simple, direct, attachment-related address specification, such as user@host.network

Once the network is accessed, it is expected that a message will go directly to the host and that the host will resolve the user name, placing the message in the user's mailbox.

3.1.4.4
Abbreviated Domain Specification

Since any number of levels is possible within the domain hierarchy, the specification of a fully qualified address can become inconvenient. RFC 822 permits abbreviated domain specification in a special case:

- For the address of the sender, call the left-most sub-domain Level N. In a header address, if all of the sub-domains above (i.e. to the right of) Level N are the same as those of the sender, then they do not have to appear in the address specification. Otherwise, the address must be fully qualified.
- This feature is subject to approval by local sub-domains. Individual sub-domains may require their member systems, which originate mail, to provide full domain specification only. When permitted, abbreviations may be present only while the message stays within the sub-domain of the sender.
- Use of this mechanism requires the sender's sub-domain to reserve the names of all top-level domains, so that full specifications can be distinguished from abbreviated specifications.

  For example, if a sender's address is

    sender@registry-A.registry-1.organization-X

  and one recipient's address is

    recipient@registry-B.registry-1.organization-X

  and another's is

    recipient@registry-C.registry-2.organization-X

then ".registry-1.organization-X" need not be specified in the message, but "registry-C.registry-2" DOES have to be specified. That is, the first two addresses may be abbreviated, but the third address must be fully specified.

When a message crosses a domain boundary, all addresses must be specified in the full format, ending with the top-level name-domain in the right-most field. It is the responsibility of mail forwarding services to ensure that addresses conform with this requirement. In the case of abbreviated addresses, the relaying service must make the necessary expansions. It should be noted that it often is difficult for such a service to locate all occurrences of address abbreviations. For example, it will not be possible to find such abbreviations within the body of the message. The "Return-Path" field can aid recipients in recovering from these errors.

Note: When passing any portion of an addr-spec onto a process which does not interpret data according to this standard (e.g. mail protocol servers), there must be NO LWSP-chars preceding or following the at-sign or any delimiting period ("."), such as shown in the above examples, and only ONE SPACE contiguous <word>s.

3.1.4.5
Domain Terms

A domain-ref must be THE official name of a registry, network, or host. It is a symbolic reference, within a name sub-domain. At times, it is necessary to bypass standard mechanisms for resolving such references, using more primitive information, such as a network host address rather than its associated host name.

To permit such references, RFC 822 provides the domain-literal construct. Its contents must conform with the needs of the sub-domain in which it is interpreted.

Domain-literals which refer to domains within the Internet specify 32-bit Internet addresses, in four 8-bit fields noted in decimal, as described in RFC 820 "Assigned Numbers." For example:

[10.0.3.19]

Note: The use of domain literals is strongly discouraged. It is permitted only as a means of bypassing temporary system limitations, such as name tables which are not complete.

The names of "top-level" domains, and the names of domains in the Internet, are registered with the Network Information Center, SRI International, Menlo Park, California.

3.1.4.6
Domain-Dependent Local Strings

The local-part of an addr-spec in mailbox specification (i.e. the host's name for the mailbox) is understood to be whatever the receiving mail protocol server allows. For example, some systems do not understand mailbox references of the form "P. D. Q. Bach", but others do.

RFC 822 treats periods (".") as lexical separators. Hence, their presence in local-parts which are not quoted strings, is detected. However, such occurrences carry NO semantics. That is, if a local-part has periods within it, an address parser will divide the local-part into several tokens, but the tokens will be treated as one uninterpreted unit. The sequence will be re-assembled, when the address is passed outside of the system such as to a mail protocol service.

For example, the address

      First.Last@Registry.Org

is legal and does not require the local-part to be surrounded with quotation marks. (However, "First Last" DOES require quoting.) The local-part of the address, when passed outside of the mail system, within the Registry.Org domain, is "First.Last", again without quotation marks.

3.1.4.7
Balancing Local-Part and Domain

In some cases, the boundary between local-part and domain can be flexible. The local-part may be a simple string, which is used for the final determination of the recipient's mailbox. All other levels of reference are, therefore, part of the domain.

For some systems, in the case of abbreviated reference to the local subordinate sub-domains, it may be possible to specify only one reference within the domain part and place the other, subordinate name-domain references within the local-part. This would appear as

      mailbox.sub1.sub2@this-domain

Such a specification would be acceptable to address parsers which conform to RFC 733, but do not support RFC 822. While contrary to the intent of RFC 822, the form is legal.

Also, some sub-domains have a specification syntax which does not conform to RFC 822. For example

      sub-net.mailbox@sub-domain.domain

uses a different parsing sequence for local-part than for domain.

Note: As a rule, the domain specification should contain fields which are encoded according to the syntax of RFC 822 and which contain generally standardised information. The local-part specification should contain only that portion of the address which deviates from the form or intention of the domain field.

3.1.4.8
Multiple Mailboxes

An individual may have several mailboxes and wish to receive mail at whatever mailbox is convenient for the sender to access. RFC 822 does not provide a means of specifying "any member of" a list of mailboxes.

A set of individuals may wish to receive mail as a single unit (i.e. a distribution list). The <group> construct permits specification of such a list.

Recipient mailboxes are specified within the bracketed part (":" - ";"). A copy of the transmitted message is to be sent to each mailbox listed. RFC 822 does not permit recursive specification of groups within groups.

While a list must be named, it is not required that the contents of the list be included. In this case, the <address> serves only as an indication of group distribution and would appear in the form

    name:;

Some mail services may provide a group-list distribution facility, accepting a single mailbox reference, expanding it to the full distribution list, and relaying the mail to the list's members. RFC 822 provides no additional syntax for indicating such a service. Using the <group> address alternative, while listing one mailbox in it, can mean either that the mailbox reference will be expanded to a list or that there is a group with one member.

3.1.4.9
Explicit Path Specification

At times, a message originator may wish to indicate the transmission path that a message should follow. This is called source routing. The normal addressing scheme, used in an addr-spec, is carefully separated from such information; the <route> portion of a route-addr is provided for such occasions. It specifies the sequence of hosts and/or transmission services that are to be traversed. Both domain-refs and domain literals may be used. But it should be noted that the use of source routing is discouraged. Unless the sender has special need of path restriction, the choice of transmission route should be left to the mail transport service.

3.1.4.10
Reserved Addresses

It often is necessary to send mail to a site, without knowing any of its valid addresses.

RFC 822 specifies a single, reserved mailbox address (local-part) which is to be valid at each site. Mail sent to that address is to be routed to a person responsible for the site's mail system or to a person with responsibility for general site operation. The name of the reserved local-part address is

    Postmaster

so that "Postmaster@domain" is required to be valid.

This reserved local-part must be matched without sensitivity to alphabetic case, so that "POSTMASTER", "postmaster", and even "poStmASteR" is to be accepted.

3.2
RFC 821 Simple Mail Transfer Protocol

The Simple Mail Transfer Protocol (SMTP) is the most often used protocol to send messages over the Internet and is specified in RFC 821. The objective of SMTP is to transfer mail reliably and efficiently and it is independent of the particular transmission subsystem and requires only a reliable ordered data stream channel.

An important feature of SMTP is its capability to relay mail across transport service environments. A transport service provides an interprocess communication environment (IPCE). An IPCE may cover one network, several networks, or a subset of a network. It is important to realise that transport systems (or IPCEs) are not one-to-one with networks. A process can communicate directly with another process through any mutually known IPCE. Mail is an application or use of interprocess communication. Mail can be communicated between processes in different IPCEs by relaying through a process connected to two (or more) IPCEs. More specifically, mail can be relayed between hosts on different transport systems by a host on both transport systems.

The SMTP design is based on the following model of communication: as the result of a user mail request, the SMTP-sender establishes a two-way transmission channel to a SMTP-receiver. The SMTP-receiver may be either the ultimate destination or an intermediate host. SMTP commands are generated by the SMTP-sender and sent to the SMTP-receiver. SMTP replies are sent from the SMTP-receiver to the SMTP-sender in response to the commands.

Once the transmission channel is established, the SMTP-sender sends a MAIL command indicating the sender of the mail. If the SMTP-receiver can accept mail it responds with an OK reply. The SMTP-sender then sends a RCPT command identifying a recipient of the mail. If the SMTP-receiver can accept mail for that recipient it responds with an OK reply; if not, it responds with a reply rejecting that recipient (but not the whole mail transaction). The SMTP-sender and SMTP-receiver may negotiate several recipients. When the recipients have been negotiated the SMTP-sender sends the mail data, terminating with a special sequence. If the SMTP-receiver successfully processes the mail data it responds with an OK reply. The dialogue is purposely lock-step, one-at-a-time.

SMTP provides mechanisms for the transmission of mail; directly from the sending user's host to the receiving user's host when the two hosts are connected to the same transport service, or via one or more relay SMTP-servers in the other case. To be able to provide the relay capability the SMTP-server must be supplied with the name of the ultimate destination host as well as the destination mailbox name.

The argument to the MAIL command is a reverse-path, which specifies who the sender of the message was. The argument to the RCPT command is a forward-path, which specifies who the mail is for. The forward-path is a source route, while the reverse-path is a return route (which may be used to return a message to the sender when an error occurs with a relayed message).

When the same message is sent to multiple recipients the SMTP encourages the transmission of only one copy of the data for all the recipients at the same destination host.

The mail commands and replies have a rigid syntax. Replies also have a numeric code. Commands and replies are not case sensitive. That is, a command or reply word may be upper case, lower case, or any mixture of upper and lower case. Note that this is not true of mailbox user names. For some hosts the user name is case sensitive, and SMTP implementations must take care to preserve the case of user names as they appear in mailbox arguments. Host names are not case sensitive.

Commands and replies are composed of characters from the ASCII character set. When the transport service provides an 8-bit byte (octet) transmission channel, each 7-bit character is transmitted right justified in an octet with the high order bit cleared to zero.

3.2.1
The SMTP Procedures

This section presents the procedures used in SMTP in several parts. First comes the basic mail procedure defined as a mail transaction. Following this are descriptions of forwarding mail, verifying mailbox names and expanding mailing lists, sending to terminals instead of or in combination with mailboxes, and the opening and closing exchanges.

3.2.1.1
Mail

SMTP mail transactions consist of three steps. The transaction is started with a MAIL command which gives the sender identification. A series of one or more RCPT commands follows giving the receiver information. Then a DATA command gives the mail data. And finally, the end of mail data indicator confirms the transaction.

The first step in the procedure is the MAIL command. The <reverse-path> contains the source mailbox.

MAIL <SP> FROM:<reverse-path> <CRLF>

This command tells the SMTP-receiver that a new mail transaction is starting and to reset all its state tables and buffers, including any recipients or mail data. It gives the reverse-path which can be used to report errors. If accepted, the receiver-SMTP returns a 250 OK reply.

The <reverse-path> can contain more than just a mailbox. The <reverse-path> is a reverse source routing list of hosts and source mailbox. The first host in the <reverse-path> should be the host sending this command.

The second step in the procedure is the RCPT command:

RCPT <SP> TO:<forward-path> <CRLF>

This command gives a forward-path identifying one recipient. If accepted, the receiver-SMTP returns a 250 OK reply, and stores the forward-path. If the recipient is unknown the receiver-SMTP returns a 550 Failure reply. This second step of the procedure can be repeated any number of times.

The <forward-path> can contain more than just a mailbox. The <forward-path> is a source routing list of hosts and the destination mailbox. The first host in the <forward-path> should be the host receiving this command.

The third step in the procedure is the DATA command:

DATA <CRLF>

If accepted, the receiver-SMTP returns a 354 Intermediate reply and considers all succeeding lines to be the message text. When the end of text is received and stored the SMTP-receiver sends a 250 OK reply.

Since the mail data is sent on the transmission channel, the end of the mail data must be indicated so that the command-and-reply dialogue can be resumed. SMTP indicates the end of the mail data by sending a line containing only a period. A transparency procedure is used to prevent this from interfering with the user's text.

Please note that the mail data includes the memo header items such as Date, Subject, To, Cc, specified in RFC 822.

The end of mail data indicator also confirms the mail transaction and tells the receiver-SMTP to now process the stored recipients and mail data. If accepted, the receiver-SMTP returns a 250 OK reply. The DATA command should fail only if the mail transaction was incomplete (for example, no recipients) or if resources are not available.

The above procedure is an example of a mail transaction. These commands must be used only in the order discussed above.

3.2.1.2
Forwarding

There are some cases where the destination information in the <forward-path> is incorrect, but the receiver-SMTP knows the correct destination. In such cases, one of the following replies should be used to allow the sender to contact the correct destination.

251 User not local; will forward to <forward-path>

This reply indicates that the receiver-SMTP knows that the user's mailbox is on another host and indicates the correct forward-path to use it in the future. Note that either the host or user or both may be different. The receiver-SMTP who sent the 251 reply takes responsibility for delivering the message correctly.

551 User not local; please try <forward-path>

This reply indicates that the receiver-SMTP knows that the user's mailbox is on another host and indicates the correct forward-path to use. Note that either the host or user or both may be different. The receiver refuses to accept mail for this user, and the sender must either redirect the mail according to the information provided or return an error response to the originating user.

3.2.1.3
Verifying and Expanding

SMTP provides as additional features, commands to verify a user name or expand a mailing list. This is done with the VRFY and EXPN commands, which have character string arguments. For the VRFY command, the string is a user name, and the response may include the full name of the user and must include the mailbox of the user. For the EXPN command, the string identifies a mailing list, and the multiline response may include the full name of the users and must give the mailboxes on the mailing list.

"User name" is a fuzzy term and used purposely. If a host implements the VRFY or EXPN commands, then at least local mailboxes must be recognised as "user names". If a host chooses to recognise other strings as "user names" that is allowed as well.

In some hosts the distinction between a mailing list and an alias for a single mailbox is a bit fuzzy, since a common data structure may hold both types of entries, and it is possible to have mailing lists of one mailbox. If a request is made to verify a mailing list, a positive response can be given if, on receipt of a message addressed to the list, it will be delivered to everyone on the list, otherwise an error should be reported (e.g. "550 That is a mailing list, not a user"). If a request is made to expand a user name, a positive response can be formed by returning a list containing one name, or an error can be reported (e.g. "550 That is a user name, not a mailing list").

In the case of a multiline reply (normal for EXPN) exactly one mailbox is to be specified on each line of the reply. In the case of an ambiguous request, for example, "VRFY Smith", where there are two Smith's, the response must be "553 User ambiguous".

The character string arguments of the VRFY and EXPN commands cannot be further restricted due to the variety of implementations of the user name and mailbox list concepts. On some systems it may be appropriate for the argument of the EXPN command to be a file name for a file containing a mailing list, but again there is a variety of file naming conventions in the Internet.

Note that the VRFY and EXPN commands are not included in the minimum implementation, and are not required to work across relays when they are implemented.

3.2.1.4
Sending and Mailing

The main purpose of SMTP is to deliver messages to user's mailboxes. A very similar service provided by some hosts is to deliver messages to user's terminals (provided the user is active on the host). The delivery to the user's mailbox is called "mailing", the delivery to the user's terminal is called "sending". Because in many hosts the implementation of sending is nearly identical to the implementation of mailing, these two functions are combined in SMTP. However the sending commands are not included in the required minimum implementation. Users should have the ability to control the writing of messages on their terminals. Most hosts permit the users to accept or refuse such messages.

The following three commands are defined to support the sending options. These are used in the mail transaction instead of the MAIL command and inform the receiver-SMTP of the special semantics of this transaction:

SEND <SP> FROM:<reverse-path> <CRLF>

The SEND command requires that the mail data be delivered to the user's terminal. If the user is not active (or not accepting terminal messages) on the host a 450 reply may be returned to a RCPT command. The mail transaction is successful if the message is delivered the terminal.

SOML <SP> FROM:<reverse-path> <CRLF>

The Send Or MaiL command requires that the mail data is delivered to the user's terminal if the user is active (and accepting terminal messages) on the host. If the user is not active (or not accepting terminal messages), then the mail data is entered into the user's mailbox. The mail transaction is successful if the message is delivered either to the terminal or the mailbox.

SAML <SP> FROM:<reverse-path> <CRLF>

The Send And MaiL command requires that the mail data is delivered to the user's terminal if the user is active (and accepting terminal messages) on the host. In any case the mail data is entered into the user's mailbox. The mail transaction is successful if the message is delivered to the mailbox.

The same reply codes that are used for the MAIL commands are used for these commands.

3.2.1.5
Opening and Closing

At the time the transmission channel is opened there is an exchange to ensure that the hosts are communicating with the hosts they think they are.

The following two commands are used in transmission channel opening and closing:

> HELO <SP> <domain> <CRLF>

> QUIT <CRLF>

In the HELO command the host sending the command identifies itself; the command may be interpreted as saying "Hello, I am <domain>".

3.2.1.6
Relaying

The forward-path may be a source route of the form

> "@ONE,@TWO:JOE@THREE",

where ONE, TWO, and THREE are hosts. This form is used to emphasise the distinction between an address and a route. The mailbox is an absolute address, and the route is information about how to get there. The two concepts should not be confused.

Conceptually the elements of the forward-path are moved to the reverse-path as the message is relayed from one server-SMTP to another. The reverse-path is a reverse source route (i.e. a source route from the current location of the message to the originator of the message). When a server-SMTP deletes its identifier from the forward-path and inserts it into the reverse-path, it must use the name it is known by in the environment it is sending into, not the environment the mail came from, in case the server-SMTP is known by different names in different environments.

If the first element of the forward-path is not the identifier of a server-SMTP a message arrives at, this element is not deleted from the forward-path but is used to determine the next server-SMTP to send the message to. In any case, the server-SMTP adds its own identifier to the reverse-path.

Using source routing the receiver-SMTP receives mail to be relayed to another server-SMTP. The receiver-SMTP may accept or reject the task of relaying the mail in the same way it accepts or rejects mail for a local user. The receiver-SMTP transforms the command arguments by moving its own identifier from the forward-path to the beginning of the reverse-path. The receiver-SMTP then becomes a sender-SMTP, establishes a transmission channel to the next SMTP in the forward-path, and sends it the mail.

The first host in the reverse-path should be the host sending the SMTP commands, and the first host in the forward-path should be the host receiving the SMTP commands.

Notice that the forward-path and reverse-path appear in the SMTP commands and replies, but not necessarily in the message. That is, there is no need for these paths and especially this syntax to appear in the "To:" , "From:", "Cc:", etc. fields of the message header.

If a server-SMTP has accepted the task of relaying the mail and later finds that the forward-path is incorrect or that the mail cannot be delivered for whatever reason, then it must construct an "undeliverable mail" notification message and send it to the originator of the undeliverable mail (as indicated by the reverse-path).

This notification message must be from the server-SMTP at this host. Of course, server-SMTPs should not send notification messages about problems with notification messages. One way to prevent loops in error reporting is to specify a null reverse-path in the MAIL command of a notification message. When such a message is relayed it is permissible to leave the reverse-path null.

3.2.1.7
Domains

Domains were introduced in the Internet in the early 80s. The use of domains changes the address space from a flat global space of simple character string host names to a hierarchically structured rooted tree of global addresses. The host name is replaced by a domain and host designator which is a sequence of domain element strings separated by periods with the understanding that the domain elements are ordered from the most specific to the most general.

Whenever domain names are used in SMTP, only the official names are used, the use of nicknames or aliases is not allowed.

3.2.1.8
Changing Rules

The TURN command may be used to reverse the roles of the two programmes communicating over the transmission channel. If programme-A is currently the sender-SMTP and it sends the TURN command and receives an ok reply (250), then programme-A becomes the receiver-SMTP.

If programme-B is currently the receiver-SMTP and it receives the TURN command and sends an ok reply (250), then programme-B becomes the sender-SMTP.

To refuse to change roles the receiver sends the 502 reply.

Please note that this command is optional. It would not normally be used in situations where the transmission channel is Transmission Control Protocol (TCP). However, when the cost of establishing the transmission channel is high, this command may be quite useful. For example, this command may be useful in supporting mail exchange by using the public switched telephone system as a transmission channel, especially if some hosts poll other hosts for mail exchanges.

3.2.2
The SMTP Specifications

3.2.2.1
SMTP Commands

Command Semantics. The SMTP commands define the mail transfer or the mail system function requested by the user. SMTP commands are character strings terminated by <CRLF>. The command codes themselves are alphabetic characters terminated by <SP> if parameters follow and <CRLF> otherwise. The syntax of mailboxes must conform to receiver site conventions.

A mail transaction involves several data objects which are transmitted as arguments to different commands. The reverse-path is the argument of the MAIL command, the forward-path is the argument of the RCPT command, and the mail data is the argument of the DATA command. These arguments or data objects must be transmitted and held pending the confirmation communicated by the end of mail data indication which finalises the transaction. The model for this is that distinct buffers are provided to hold the types of data objects, that is, there is a reverse-path buffer, a forward-path buffer, and a mail data buffer. Specific commands cause information to be appended to a specific buffer, or cause one or more buffers to be cleared.

HELLO (HELO).
This command is used to identify the sender-SMTP to the receiver-SMTP. The argument field contains the host name of the sender-SMTP.

The receiver-SMTP identifies itself to the sender-SMTP in the connection greeting reply, and in the response to this command.

This command and an OK reply to it confirm that both the sender-SMTP and the receiver-SMTP are in the initial state, that is, there is no transaction in progress and all state tables and buffers are cleared.

MAIL (MAIL).

This command is used to initiate a mail transaction in which the mail data is delivered to one or more mailboxes. The argument field contains a reverse-path.

The reverse-path consists of an optional list of hosts and the sender mailbox. When the list of hosts is present, it is a "reverse" source route and indicates that the mail was relayed through each host on the list (the first host in the list was the most recent relay). This list is used as a source route to return non-delivery notices to the sender. As each relay host adds itself to the beginning of the list, it must use its name as known in the IPCE to which it is relaying the mail rather than the IPCE from which the mail came (if they are different). In some types of error reporting messages (for example, undeliverable mail notifications) the reverse-path may be null.

This command clears the reverse-path buffer, the forward-path buffer and the mail data buffer, and inserts the reverse-path information from this command into the reverse-path buffer.

RECIPIENT (RCPT).

This command is used to identify an individual recipient of the mail data; multiple recipients are specified by multiple use of this command.

The forward-path consists of an optional list of hosts and a required destination mailbox. When the list of hosts is present, it is a source route and indicates that the mail must be relayed to the next host on the list. If the receiver-SMTP does not implement the relay function, it may use the same reply it would do for an unknown local user (550).

When mail is relayed, the relay host must remove itself from the beginning forward-path and put itself at the beginning of the reverse-path. When mail reaches its ultimate destination (the forward-path contains only a destination mailbox), the receiver-SMTP inserts it into the destination mailbox in accordance with its host mail conventions.

For example, mail received at relay host A with arguments

        FROM:<USERX@HOSTY.ARPA>
        TO:<@HOSTA.ARPA,@HOSTB.ARPA:USERC@HOSTD.ARPA>

will be relayed on to host B with arguments

        FROM:<@HOSTA.ARPA:USERX@HOSTY.ARPA>
        TO:<@HOSTB.ARPA:USERC@HOSTD.ARPA>.

This command causes its forward-path argument to be appended to the forward-path buffer.

DATA (DATA)

The receiver treats the lines following this command as mail data from the sender. This command causes the sent mail data to be appended to the mail data buffer. The mail data may contain any of the 128 ASCII character codes.

The mail data is terminated by a line containing only a period, that is the character sequence "<CRLF>.<CRLF>". This is the end of mail data indication.

The end of mail data indication requires that the receiver-SMTP must now process the stored mail transaction information. This processing consumes the information in the reverse-path buffer, the forward-path buffer, and the mail data buffer, and after completion of this command, these buffers are cleared. If the processing is successful, the receiver-SMTP must send an OK reply. If the processing fails completely, the receiver-SMTP must send a failure reply.

When the receiver-SMTP accepts a message either for relaying or for final delivery, it inserts a time stamp line at the beginning of the mail data. The time stamp line indicates the identity of the host that sent the message, and the identity of the host that received the message and the date and time the message was received. Relayed messages will have multiple time stamp lines.

When the receiver-SMTP makes the "final delivery" of a message, it inserts a return path line at the beginning of the mail data. The return path line preserves the information in the <reverse-path> from the MAIL command. Here, final delivery means the message leaves the SMTP world. Normally, this would mean it has been delivered to the destination user, but in some cases it may be further processed and transmitted by another mail system.

It is possible for the mailbox in the return path to be different from the actual sender's mailbox, for example, if error responses are to be delivered, a special error handling mailbox, rather than the message sender's mailbox, may be present.

The preceding two paragraphs imply that the final mail data will begin with a return path line, followed by one or more time stamp lines. These lines will be followed by the mail data header and body conformant to RFC 822.

Special mention is needed regarding the response and further action required when the processing following the end of mail data indication is only partially successful. This could arise if, after accepting several recipients and the mail data, the receiver-SMTP finds that the mail data can be successfully delivered to some of the recipients, but it cannot be to others (for example, due to mailbox space allocation problems). In such a situation, the response to the DATA command must be an OK reply. But the receiver-SMTP must compose and send an "undeliverable mail" notification message to the originator of the message. Either a single notification which lists all of the recipients that failed to get the message, or separate notification messages must be sent for each failed recipient. All undeliverable mail notification messages are sent using the MAIL command (even if they result from processing a SEND, SOML, or SAML command).

SEND (SEND)

This command is used to initiate a mail transaction in which the mail data is delivered to one or more terminals. The argument field contains a reverse-path. This command is successful if the message is delivered to a terminal.

The reverse-path consists of an optional list of hosts and the sender mailbox. When the list of hosts is present, it is a "reverse" source route and indicates that the mail was relayed through each host on the list (the first host in the list was the most recent relay). This list is used as a source route to return non-delivery notices to the sender. As each relay host adds itself to the beginning of the list, it must use its name as known in the IPCE to which it is relaying the mail rather than the IPCE from which the mail came (if they are different).

This command clears the reverse-path buffer, the forward-path buffer and the mail data buffer, and inserts the reverse-path information from this command into the reverse-path buffer.

SEND OR MAIL (SOML)

This command is used to initiate a mail transaction in which the mail data is delivered to one or more terminals or mailboxes. For each recipient the mail data is delivered to the recipient's terminal if the recipient is active on the host (and accepting terminal messages), otherwise it is delivered to the recipient's mailbox. The argument field contains a reverse-path. This command is successful if the message is delivered to a terminal or the mailbox.

For the reverse-path the same rules apply as to the reverse-path of the SEND command.

This command clears the reverse-path buffer, the forward-path buffer and the mail data buffer, and inserts the reverse-path information from this command into the reverse-path buffer.

SEND AND MAIL (SAML)

This command is used to initiate a mail transaction in which the mail data is delivered to one or more terminals and mailboxes. For each recipient the mail data is delivered to the recipient's terminal if the recipient is active on the host (and accepting terminal messages), and for all recipients to the recipient's mailbox. The argument field contains a reverse-path. This command is successful if the message is delivered to the mailbox.

For the reverse-path the same rules apply as to the reverse-path of the SEND command.

This command clears the reverse-path buffer, the forward-path buffer and the mail data buffer, and inserts the reverse-path information from this command into the reverse-path buffer.

RESET (RSET)

This command specifies that the current mail transaction is to be aborted. Any stored sender, recipients, and mail data must be discarded, and all buffers and state tables cleared.

The receiver must send an OK reply.

VERIFY (VRFY)

This command asks the receiver to confirm that the argument identifies a user. If it is a user name, the full name of the user (if known) and the fully specified mailbox are returned.

This command has no effect on any of the reverse-path buffer, the forward-path buffer, or the mail data buffer.

EXPAND (EXPN)

This command asks the receiver to confirm that the argument identifies a mailing list, and if so, to return the membership of that list. The full name of the users (if known) and the fully specified mailboxes are returned in a multiline reply.

This command has no effect on any of the reverse-path buffer, the forward-path buffer, or the mail data buffer.

HELP (HELP)

This command causes the receiver to send helpful information to the sender of the HELP command. The command may take an argument (e.g. any command name) and return more specific information as a response.

This command has no effect on any of the reverse-path buffer, the forward-path buffer, or the mail data buffer.

NOOP (NOOP)

This command does not affect any parameters or previously entered commands. It specifies no action other than that the receiver send an OK reply.

This command has no effect on any of the reverse-path buffer, the forward-path buffer, or the mail data buffer.

QUIT (QUIT)

This command specifies that the receiver must send an OK reply, and then close the transmission channel.

The receiver should not close the transmission channel until it receives and replies to a QUIT command (even if there was an error). The sender should not close the transmission channel until it sends a QUIT command and receives the reply (even if there was an error response to a previous command). If the connection is closed prematurely the receiver should act as if a RSET command had been received (cancelling any pending transaction, but not undoing any previously completed transaction), the sender should act as if the command or transaction in progress had received a temporary error (4xx).

TURN (TURN)

This command specifies that the receiver-SMTP must either

- send an OK reply and then take on the role of the sender-SMTP, or
- send a refusal reply and retain the role of the receiver-SMTP.

If programme-A is currently the sender-SMTP and it sends the TURN command and receives an OK reply (250), then programme-A becomes the receiver-SMTP. Programme-A is then in the initial state as if the transmission channel just opened, and it then sends the 220 service ready greeting.

If programme-B is currently the receiver-SMTP and it receives the TURN command and sends an OK reply (250), then programme-B becomes the sender-SMTP. Programme-B is then in the initial state as if the transmission channel just opened, and it then expects to receive the 220 service ready greeting. To refuse to change roles the receiver sends the 502 reply.

There are restrictions on the order in which the SMTP commands may be used. The first command in a session must be the HELO command. The HELO command may be used later in a session as well. If the HELO command argument is not acceptable a 501 failure reply must be returned and the SMTP-receiver must stay in the same state. In the case where the extension to SMTP are requested the first command is not the HELO command but the EHLO command. This additional command indicates whether the extensions are available or not. If the SMTP-server supports the extensions it will give a successful response, a failure response or an error response. If the SMTP-server does not support any SMTP service extensions it will generate an error response.

The NOOP, HELP, EXPN, and VRFY commands can be used at any time during a session.

The MAIL, SEND, SOML, or SAML commands begin a mail transaction. Once started a mail transaction consists of one of the transaction beginning commands, one or more RCPT commands, and a DATA command, in that order. A mail transaction may be aborted by the RSET command. There may be zero or more transactions in one session.

The last command in a session must be the QUIT command. The QUIT command can not be used at any other time in a session.

SMTP REPLIES

Replies to SMTP commands are devised to ensure the synchronisation of requests and actions in the process of mail transfer, and to guarantee that the sender-SMTP always knows the state of the receiver-SMTP. Every command must generate exactly one reply.

An SMTP reply consists of a three digit number (transmitted as three alphanumeric characters) followed by some text. The number is intended for use by automata to determine what state to enter next; the text is meant for the human user. It is intended that the three digits contain enough encoded information that the sender-SMTP need not examine the text and may either discard it or pass it on to the user, as appropriate. In particular, the text may be receiver-dependent and context-dependent, so there are likely to be varying texts for each reply code. Formally, a reply is defined to be the sequence:

- a three-digit code, <SP>
- one line of text, and <CRLF>,
- or a multiline reply.

Only the EXPN and HELP commands are expected to result in multiline replies in normal circumstances, however, multiline replies are allowed for any command.

A more detailed explanation of the reply codes and the theory how they are generated can be found in RFC 821.

3.2.2.2
Numeric Order List of Reply Codes

211 System status, or system help reply

214 Help message
   [Information on how to use the receiver or the meaning of a particular nonstandard command; this reply is useful only to the human user]

220 <domain> Service ready

221 <domain> Service closing transmission channel

250 Requested mail action okay, completed

251 User not local; will forward to <forward-path>

354 Start mail input; end with <CRLF>.<CRLF>

421 <domain> Service not available, closing transmission channel
   [This may be a reply to any command if the service knows it must shut down]

450 Requested mail action not taken: mailbox unavailable [E.g. mailbox busy]

451 Requested action aborted: local error in processing

452 Requested action not taken: insufficient system storage

500 Syntax error, command unrecognised
   [This may include errors such as command line too long]

501 Syntax error in parameters or arguments

502 Command not implemented

503 Bad sequence of commands

504 Command parameter not implemented

550 Requested action not taken: mailbox unavailable

[E.g. mailbox not found, no access]
551 User not local; please try <forward-path>
552 Requested mail action aborted: exceeded storage allocation
553 Requested action not taken: mailbox name not allowed
[E.g. mailbox syntax incorrect]
554 Transaction failed

3.3
SMTP Service Extensions

SMTP has provided a stable, effective basis for the relay function of MTAs. Although more than 15 years old, SMTP has proven remarkably resilient. Nevertheless, the need for a number of protocol extensions has become evident and therefore in 1994 the first SMTP service extensions were specified in RFC 1651. Additionally many other RFCs are dealing with SMTP service extensions. Rather than describing these extensions as separate and haphazard entities, RFC 1651 (and its successor RFC 1869 published in 1996) enhances SMTP in a straightforward fashion that provides a framework in which all future extensions can be built in a single consistent way.

3.3.1
Framework for SMTP Extensions

For the purpose of service extensions to SMTP, SMTP relays a mail object containing an envelope and a content and the following rules apply:

* The SMTP envelope is straightforward, and is sent as a series of SMTP protocol units: it consists of an originator address (to which error reports should be directed); a delivery mode (e.g. deliver to recipient mailboxes); and one or more recipient addresses.
* The SMTP content is sent in the SMTP DATA protocol unit and has two parts: the headers and the body. The headers form a collection of field/value pairs structured according to RFC 822, whilst the body, if structured, is defined according to the Multipurpose Internet Mail Extensions (MIME). The content is textual in nature, expressed using the US-ASCII repertoire (ANSI X3.4-1986). Although extensions (such as MIME) may relax this restriction for the content body, the content headers are always encoded using the US-ASCII repertoire.
* The algorithm defined in RFC 1522 is used to represent header values outside the US-ASCII repertoire, whilst still encoding them using the US-ASCII repertoire.

Although SMTP is widely and robustly deployed, experience has shown that there are some SMTP service extensions required. RFC 1869 defines a means whereby both an extended SMTP client and server may recognise each other as such and the server can inform the client as to the service extensions that it supports.

It must be emphasised that any extension to the SMTP service should not be considered lightly. SMTP's strength comes primarily from its simplicity. It is very important for not losing this property that each and every extension, regardless of its benefits, must be carefully scrutinised with respect to its implementation, deployment, and interoperability costs. In many cases, the cost of extending the SMTP service will likely outweigh the benefit.

Given this environment, the framework for the extensions described in RFC 1651 consists of

- a new SMTP command (EHLO);
- a registry of SMTP service extensions;
- additional parameters to the SMTP MAIL FROM and RCPT TO commands.

3.3.2
The EHLO Command

A client SMTP supporting SMTP service extensions should start an SMTP session by issuing the EHLO command instead of the HELO command. If the SMTP server supports the SMTP service extensions, it will give

- a successful response;
- a failure response;
- an error response.

If the SMTP server does not support any SMTP service extensions it will generate an error response.

3.3.2.1
Required Changes to RFC 821

RFC 1869 is intended to extend RFC 821 without impacting existing services in any way. RFC 821 states that the first command in an SMTP session must be the HELO command. This requirement is amended to allow a session to start with either EHLO or HELO. Additionally RFC 1869 extends the SMTP MAIL FROM and RCPT TO commands to allow additional parameters and parameter values. RFC 1869 amended the restriction for the length of the command lines for both commands to 512 characters. The limit still exists for command lines without parameters. Furthermore it is specified that all future specification of the MAIL FROM and RCPT TO commands must also specify the maximal length for each parameter value. The maximal command length for any SMTP implementation with extensions is 512 plus the sum of all the maximum parameter lengths for all supported extensions.

3.3.2.2
Command Syntax

The syntax for this command, using the Augmented Backus-Naur Form (ABNF) notation of RFC 822, is

ehlo-cmd ::= "EHLO" SP domain CR LF

If successful, the server SMTP responds with code 250. On failure, the server SMTP responds with code 550. On error, the server SMTP responds with one of codes 500, 501, 502, 504, or 421.

This command is issued instead of the HELO command and may be issued at any time that a HELO command would be appropriate. That is, if the EHLO command is issued, and a successful response is returned, then a subsequent HELO or EHLO command will result in the server SMTP replying with code 503. A client SMTP must not cache any information returned if the EHLO command succeeds. That is, a client SMTP must issue the EHLO command at the start of each SMTP session if information about extended facilities is needed.

3.3.2.3
Successful Response

If the server-SMTP implements and is able to perform the EHLO command, it will return code 250. This indicates that both the server and client SMTP are in the initial state, that is, there is no transaction in progress and all state tables and buffers are cleared.

Normally, this response will be a multiline reply. Each line of the response contains a keyword and, optionally, one or more parameters.

The syntax for a positive response, using the ABNF notation of RFC 822 is

ehlo-ok-rsp ::= "250" domain [SP greeting] CR LF   the usual HELO chit-chat
/ ("250-" domain [SP greeting] CR LF
*("250-" ehlo-line CR LF)
"250" SP ehlo-line CR LF)

greeting ::= 1*<any character other than CR or LF>

ehlo-line ::= ehlo-keyword *(SP ehlo-param)

ehlo-keyword ::= (ALPHA / DIGIT) *(ALPHA / DIGIT / "-")
syntax and values depend on ehlo-keyword

ehlo-param ::= 1*<any CHAR excluding SP and all control characters (US-ASCII 0-31 inclusive)>

ALPHA ::= <any one of the 52 alphabetic characters (A through Z in upper
            case, and, a through z in lower case)>

DIGIT ::= <any one of the 10 numeric characters (0 through 9)>

CR ::= <the carriage-return character (ASCII decimal code 13)>

LF ::= <the line-feed character (ASCII decimal code 10)>

SP ::= <the space character (ASCII decimal code 32)>

Although EHLO keywords may be specified in upper, lower, or mixed case,
they must always be recognised and processed in a case-insensitive manner. This
is simply an extension of practices begun in RFC 821.

The Internet Assigned Numbers Authority (IANA) maintains a registry of
standard SMTP service extensions. Associated with each such extension is a
corresponding EHLO keyword value. Each service extension registered with the
IANA is defined by a standards-track RFC, and such a definition includes

- the textual name of the SMTP service extension;
- the EHLO keyword value associated with the extension;
- the syntax and possible values of parameters associated with the EHLO
  keyword value;
- any additional SMTP verbs associated with the extension (additional verbs
  will usually be, but are not required to be, the same as the EHLO keyword
  value);
- any new parameters the extension associates with the MAIL FROM or RCPT
  TO verbs; and,
- how support for the extension affects the behaviour of a server and client
  SMTP;
- the increment by which the extension is increasing the maximum length of the
  commands MAIL FROM and RCPT TO, or both, over that specified in RFC
  821.

In addition, any EHLO keyword value that starts with an upper or lower case
"X" refers to a local SMTP service extension, which is used through bilateral,
rather than standardised, agreement. Keywords beginning with "X" may not be
used in a registered service extension.

Any keyword values presented in the EHLO response that do not begin with
"X" must correspond to a standard, standards-track or Internet Engineering
Steering Group (IESG)-approved experimental SMTP service extension
registered with IANA. A conforming server must not offer non-"X" prefixed
keyword values that are not described in a registered and standardised extension.

Additional verbs are bound by the same rules as EHLO keywords.
Specifically, verbs beginning with "X" are local extensions that may not be
standardised and verbs not beginning with "X" must always be registered.

3.3.2.4
Failure Response

If for some reason the server SMTP is unable to list the service extensions it supports, it will return code 554. In the case of a failure response, the client SMTP should issue either the HELO or QUIT command.

3.3.2.5
Error Responses from Extended Servers

If the server SMTP recognises the EHLO command, but the command argument is unacceptable, it will return code 501.

If the server SMTP recognises, but does not implement, the EHLO command, it will return code 502.

If the server SMTP determines that the SMTP service is no longer available (e.g. due to imminent system shutdown), it will return code 421.

In the case of any error response, the client SMTP should issue either the HELO or QUIT command.

3.3.2.6
Responses from Servers Without Extensions

A server-SMTP that conforms to RFC 821 but does not support the extensions will not recognise the EHLO command and will consequently return code 500, as specified in RFC 821. The server SMTP should stay in the same state after returning this code (see section 4.1.1 of RFC 821). The client SMTP may then issue either a HELO or a QUIT command.

3.3.2.7
Responses from Improperly Implemented Servers

Some SMTP servers are known to disconnect the SMTP transmission channel upon receipt of the EHLO command. The disconnect can occur immediately or after sending a response. Such behaviour violates section 4.1.1 of RFC 821, which explicitly states that disconnection should only occur after a QUIT command is issued.

Nevertheless, in order to achieve maxmimum interoperability, it is suggested that extended SMTP clients using EHLO be coded to check for server connection closure after EHLO is sent, either before or after returning a reply. If this happens the client must decide if the operation can be successfully completed without using any SMTP extensions. If it can a new connection can be opened and the HELO command can be used.

Other improperly implemented servers will not accept a HELO command after EHLO has been sent and rejected. In some cases, this problem can be worked around by sending a RSET after the failure response to EHLO, then sending the HELO. Clients that do this should be aware that many implementations will return a failure code (e.g. 503 Bad sequence of commands) in response to the RSET. This code can be safely ignored.

3.3.3
Initial IANA Registry

The IANA's initial registry of SMTP service extensions consists of the following entries:

Service Ext	EHLO Keyword	Parameters	Verb	Added Behaviour
Send SEND	none	SEND	defined in RFC 821	
Send or Mail	SOML	none	SOML	defined in RFC 821
Send and Mail	SAML	none	SAML	defined in RFC 821
Expand	EXPN	none	EXPN	defined in RFC 821
Help HELP	none	HELP	defined in RFC 821	
Turn TURN	none	TURN	defined in RFC 821	

These commands are defined in RFC 821 as optional in contrast to the following commands that are defined as mandatory and have therefore to be present in any minimal SMTP implementation: HELO, MAIL, RCPT, DATA, RSET, VRFY, NOOP, and QUIT.

3.3.4
MAIL FROM and RCPT TO Parameters

It is recognised that several of the extensions planned for SMTP will make use of additional parameters associated with the MAIL FROM and RCPT TO command. The syntax for these commands, again using the ABNF notation of RFC 822 as well as underlying definitions from RFC 821, is

esmtp-cmd ::= inner-esmtp-cmd [SP esmtp-parameters] CR LF

esmtp-parameters ::= esmtp-parameter *(SP esmtp-parameter)

esmtp-parameter ::= esmtp-keyword ["=" esmtp-value]

esmtp-keyword ::= (ALPHA / DIGIT) *(ALPHA / DIGIT / "-")
            syntax and values depend on esmtp-keyword

esmtp-value ::= 1*<any CHAR excluding "=", SP, and all control characters (US-ASCII 0-31 inclusive)>
            The following commands are extended to accept extended parameters.

inner-esmtp-cmd   ::=   ("MAIL   FROM:<"   reverse-path   ">")   /   ("RCPT
TO:<"forward-path">")

All esmtp-keyword values must be registered as part of the IANA registration process described above. This definition only provides the framework for future extension; no extended MAIL FROM or RCPT TO parameters are defined by either RFC 1651 or RFC 1869.

3.3.4.1
Error Responses

If the server SMTP does not recognise or cannot implement one or more of the parameters associated with a particular MAIL FROM or RCPT TO command, it will return code 555.

If for some reason the server is temporarily unable to accommodate one or more of the parameters associated with a MAIL FROM or RCPT TO command, and if the definition of the specific parameter does not mandate the use of another code, it should return code 455.

Errors specific to particular parameters and their values will be specified in the parameter's defining RFC.

3.3.5
Received: Header Field Annotation

SMTP servers are required to add an appropriate Received: field to the headers of all messages they receive. A "with ESMTP" clause should be added to this field when any of the SMTP service extensions is used. "ESMTP" is hereby added to the list of standard protocol names registered with IANA.

3.4
Delivery Status Notifications (DSN)

The SMTP protocol requires that an SMTP server provides notification of delivery failure, if it determines that a message cannot be delivered to one or more recipients. Traditionally, such notifications consist of an ordinary Internet mail message with a format specified in RFC 822, sent back to the originator of the message (the argument of the SMTP MAIL command), containing an explanation of the error and at least the headers of the failed message.

However, there are cases where such messages are insufficient to diagnose problems, or even to determine at which host or for which recipients a problem occurred. Additionally, for the exchange of such notifications with other message handling systems, a standardised format for delivery notifications in Internet mail is required, which

- is reliable, in the sense that any Delivery Status Notification (DSN) request will either be honoured at the time of final delivery, or result in a response that indicates that the request cannot be honoured;
- when both success and failure notifications are requested, provides an unambiguous and nonconflicting indication of whether delivery of a message to a recipient succeeded or failed,
- is stable, in that a failed attempt to deliver a DSN should never result in the transmission of another DSN over the network,
- preserves sufficient information to allow the sender to identify both the mail transaction and the recipient address which caused the notification, even when mail is forwarded or gatewayed to foreign environments, and
- interfaces acceptably with non-SMTP and non-rfc822-based mail systems, both so that notifications returned from foreign mail systems may be useful to Internet users, and so that the notification requests from foreign environments may be honoured. Among the requirements implied by this goal are the ability to request non-return-of-content, and the ability to specify whether positive delivery notifications, negative delivery notifications, both, or neither, should be issued.

In order to meet this requirement in 1996 the SMTP service extensions for DSNs were specified in RFC 1891 and their format in RFC 1894. RFC 1891 specifies the requirements for MTAs to be able to handle DSN.

3.4.1
Framework for the Delivery Status Notifications

The following service extension is therefore defined:

- The name of the SMTP service extension is "Delivery Status Notification";
- the EHLO keyword value associated with this extension is "DSN";
- no parameters are allowed with this EHLO keyword value;
- two optional parameters are added to the RCPT command (NOTIFY, ORCPT) and two optional parameters are added to the MAIL command (RET, ENVID);
- no additional SMTP verbs are defined.

3.4.2
The Delivery Status Notification Service Extension

An SMTP client wishing to request a DSN for a message may issue the EHLO command to start an SMTP session, to determine if the server supports any of several service extensions. If the server responds with code 250 to the EHLO command, and the response includes the EHLO keyword DSN, then the DSN extension is supported.

Ordinarily, when an SMTP server returns a positive (2xx) reply code in response to a RCPT command, it agrees to accept responsibility for either delivering the message to the named recipient, or sending a notification to the sender of the message indicating that delivery has failed. However, an extended SMTP ("ESMTP") server which implements this service extension will accept an optional NOTIFY parameter with the RCPT command. If present, the NOTIFY parameter alters the conditions for generation of delivery status notifications from the default (issue notifications only on failure) as specified in RFC 821. The ESMTP client may also request (via the RET parameter) whether the entire contents of the original message should be returned (as opposed to just the headers of that message) along with the DSN.

In general, an ESMTP server which implements this service extension will propagate DSN requests when relaying mail to other SMTP-based MTAs which also support this extension, and make a "best effort" to ensure that such requests are honoured when messages are passed into other environments.

In order that any DSNs thus generated will be meaningful to the sender, any ESMTP server which supports this extension will attempt to propagate the following information to any other MTAs that are used to relay the message, for use in generating DSNs:

- for each recipient, a copy of the original recipient address, as used by the sender of the message.
  This address need not be the same as the mailbox specified in the RCPT command. For example, if a message was originally addressed to A@B.C and later forwarded to A@D.E, after such forwarding has taken place, the RCPT command will specify a mailbox of A@D.E. However, the original recipient address remains A@B.C.
  Also, if the message originated from an environment which does not use Internet-style user@domain addresses, and was gatewayed into SMTP, the original recipient address will preserve the original form of the recipient address.
- for the entire SMTP transaction, an envelope identification string, which may be used by the sender to associate any DSNs with the transaction used to send the original message.

3.4.2.1
Additional Parameters for RCPT and MAIL Commands

The extended RCPT and MAIL commands are issued by a client when it wishes to request a DSN from the server, under certain conditions, for a particular recipient. The extended RCPT and MAIL commands are identical to the RCPT and MAIL commands defined in RFC 821, except that one or more of the following parameters appear after the sender or recipient address, respectively.

3.4.2.2
The NOTIFY Parameter of the ESMTP RCPT Command

A RCPT command issued by a client may contain the optional esmtp-keyword "NOTIFY", to specify the conditions under which the SMTP server should generate DSNs for that recipient. If the NOTIFY esmtp-keyword is used, it must have an associated esmtp-value, formatted according to the following rules, using the ABNF of RFC 822:

notify-esmtp-value = "NEVER" / 1#notify-list-element

notify-list-element = "SUCCESS" / "FAILURE" / "DELAY"

Notes:

- Multiple notify-list-elements, separated by commas, may appear in a NOTIFY parameter. The only keyword that must appear unaccompanied by any other is the NEVER keyword.
- Any of the keywords NEVER, SUCCESS, FAILURE, or DELAY may be spelled in any combination of upper and lower case letters.

The meaning of the NOTIFY parameter values is generally as follows:

- A NOTIFY parameter value of "NEVER" requests that a DSN not be returned to the sender under any conditions.
- A NOTIFY parameter value containing the "SUCCESS" or "FAILURE" keywords requests that a DSN be issued on successful delivery or delivery failure, respectively.
- A NOTIFY parameter value containing the keyword "DELAY" indicates the sender's willingness to receive "delayed" DSNs. Delayed DSNs may be issued if delivery of a message has been delayed for an unusual amount of time (as determined by the MTA at which the message is delayed), but the final delivery status (whether successful or failure) cannot be determined. The absence of the DELAY keyword in a NOTIFY parameter requests that a "delayed" DSN not be issued under any condition.

For compatibility with SMTP clients that do not use the NOTIFY facility, the absence of a NOTIFY parameter in a RCPT command may be interpreted as either NOTIFY=FAILURE or NOTIFY=FAILURE,DELAY.

3.4.2.3
The ORCPT Parameter of the ESMTP RCPT Command

The ORCPT esmtp-keyword of the RCPT command is used to specify an "original" recipient address that corresponds to the actual recipient to which the message is to be delivered. If the ORCPT esmtp-keyword is used, it must have an associated esmtp-value, which consists of the original recipient address, encoded according to the rules below.

The ABNF for the ORCPT parameter is

orcpt-parameter = "ORCPT=" original-recipient-address

original-recipient-address = addr-type ";" xtext

addr-type = atom

The "addr-type" portion must be an IANA-registered electronic mail address-type (as defined in RFC 1894), while the "xtext" portion contains an encoded representation of the original recipient address using the rules of RFC 1891. The entire ORCPT parameter may be up to 500 characters in length.

When initially submitting a message via SMTP, if the ORCPT parameter is used, it must contain the same address as the RCPT TO address (unlike the RCPT TO address, the ORCPT parameter will be encoded as xtext). Likewise, when a mailing list submits a message via SMTP to be distributed to the list subscribers, if ORCPT is used, the ORCPT parameter must match the new RCPT TO address of each recipient, not the address specified by the original sender of the message.

The "addr-type" portion of the original-recipient-address is used to indicate the "type" of the address which appears in the ORCPT parameter value. However, the address associated with the ORCPT keyword is not constrained to conform to the syntax rules for that "addr-type".

Ideally, the "xtext" portion of the original-recipient-address should contain, in encoded form, the same sequence of characters that the sender used to specify the recipient. However, for a message gatewayed from an environment (such as X.400) in which a recipient address is not a simple string of printable characters, the representation of recipient address must be defined by a specification for gatewaying between DSNs and that environment.

3.4.2.4
The RET Parameter of the ESMTP MAIL Command

The RET esmtp-keyword on the extended MAIL command specifies whether or not the message should be included in any failed DSN issued for this message transmission. If the RET esmtp-keyword is used, it must have an associated esmtp-value, which is one of the following keywords:

- FULL requests that the entire message be returned in any "failed "DSN issued for this recipient.
- HDRS requests that only the headers of the message be returned.

The FULL and HDRS keywords may be spelled in any combination of upper and lower case letters.

If no RET parameter is supplied, the MTA may return either the headers of the message or the entire message for any DSN containing indication of failed deliveries.

Note that the RET parameter only applies to DSNs that indicate delivery failure for at least one recipient. If a DSN contains no indications of delivery failure, only the headers of the message should be returned.

3.4.2.5
The ENVID Parameter of the ESMTP MAIL Command

The ENVID esmtp-keyword of the SMTP MAIL command is used to specify an "envelope identifier" to be transmitted along with the message and included in any DSNs issued for any of the recipients named in this SMTP transaction. The purpose of the envelope identifier is to allow the sender of a message to identify the transaction for which the DSN was issued.

The ABNF for the ENVID parameter is

envid-parameter = "ENVID=" xtext

The ENVID esmtp-keyword must have an associated esmtp-value. No meaning is assigned by the mail system to the presence or absence of this parameter or to any esmtp-value associated with this parameter; the information is used only by the sender or his user agent. The ENVID parameter may be up to 100 characters in length.

3.4.2.6
Restrictions on the Use of DSN Parameters

The RET and ENVID parameters must not appear more than once each in any single MAIL command. If more than one of either of these parameters appears in a MAIL command, the ESMTP server should respond with "501 syntax error in parameters or arguments".

The NOTIFY and ORCPT parameters must not appear more than once in any RCPT command. If more than one of either of these parameters appears in a RCPT command, the ESMTP server should respond with "501 syntax error in parameters or arguments".

3.5
Multipurpose Internet Mail Extensions (MIME)

Since its publication in 1982, RFC 822 has defined the standard format of textual mail messages on the Internet. Its success has been in a way that the RFC 822 format has been adopted, wholly or partially, well beyond the confines of the Internet and the Internet SMTP transport defined by RFC 821. As the format has seen wider use, a number of limitations have proven increasingly restrictive for the user community.

RFC 822 was intended to specify a format for text messages only. As a consequence, nontext messages, like multimedia messages that might include audio, video, or image files, are simply not mentioned. Even in the case of text, however, RFC 822 is inadequate for the needs of mail users whose languages require the use of character sets richer than US-ASCII.

Since RFC 822 does not specify mechanisms for mail containing audio, video, Asian language text, or even text in most European languages, it became clear that additional specifications are needed.

One of the notable limitations of RFC 821/822 based mail systems is the fact that they limit the contents of electronic mail messages to relatively short lines (e.g. 1000 characters or less) of 7bit US-ASCII. This forces users to convert any nontextual data that they may wish to send into seven-bit bytes representable as printable US-ASCII characters before invoking a local mail UA for sending the mail.

Examples of such encodings currently used in the Internet include pure hexadecimal, uuencode, the 3-in-4 base 64 scheme specified in RFC 1421, the Andrew Toolkit Representation, and many others.

The limitations of RFC 822 mail become even more apparent as gateways are designed to allow for the exchange of mail messages between RFC 822 hosts and X.400 hosts because X.400 does specify mechanisms for the inclusion of non-textual material within electronic mail messages.

Even though a UA may not have the capability of dealing with the nontextual material, the user might have some mechanism external to the UA that can extract useful information from the material. Moreover, it does not allow for the fact that the message may eventually be gatewayed back into an X.400 message handling system (i.e. the X.400 message is "tunnelled" through Internet mail), where the non-textual information would definitely become useful again.

In order to meet the requirements for multimedia messages several RFCs were published that deal with the extensions to RFC 822. The latest ones are RFC 2045-RFC2049. All of them describe several mechanisms that combine to solve most of these problems mentioned above without introducing any serious incompatibilities with the existing world of RFC 822 mail. In particular, they describes

- A MIME-Version header field, which uses a version number to declare a message to be conformant with MIME and allows mail-processing agents to distinguish between such messages and those generated by older or non-conformant software, which are presumed to lack such a field.
- A Content-Type header field, generalised from RFC 1049, which can be used to specify the media type and subtype of data in the body of a message and to fully specify the native representation (canonical form) of such data.

- A Content-Transfer-Encoding header field, which can be used to specify both the encoding transformation that was applied to the body and the domain of the result. Encoding transformations other than the identity transformation are usually applied to data in order to allow it to pass through mail transport mechanisms which may have data or character set limitations.
- Two additional header fields that can be used to further describe the data in a body, the Content-ID and Content-Description header fields.

3.5.1
MIME Header Fields

MIME defines a number of new RFC 822 header fields that are used to describe the content of a MIME entity. These header fields occur in at least two contexts:

- As part of a regular RFC 822 message header.
- In a MIME body part header within a multipart construct.

The formal definition of these header fields is as follows:

```
entity-headers := [content CRLF]
                  [encoding CRLF]
                  [id CRLF]
                  [description CRLF]
                  *(MIME-extension-field CRLF)
```

```
MIME-message-headers := entity-headers
                        fields
                        version CRLF
```

The ordering of the header fields implied by this BNF definition should be ignored.

```
MIME-part-headers := entity-headers
                     [fields]
```

Any field not beginning with "content-" can have no defined meaning and may be ignored. The ordering of the header fields implied by this BNF definition should be ignored.

3.5.2
MIME-Version Header Field

Since RFC 822 was published in 1982, there has been only one format standard for Internet messages, and therefore there has been little perceived need to declare the format standard in use. RFC 2045 is an independent specification that complements RFC 822. Although the extensions in RFC 2045 have been defined in a way that is compatible with RFC 822, there are still circumstances in which it might be desirable for a mail-processing agent to know whether a message was composed with the new standard in mind.

Therefore, RFC 2045 defines a new header field, "MIME-Version", which is to be used to declare the version of the Internet message body format standard in use.

Messages composed in accordance with RFC 2045 must include such a header field, with the following verbatim text:

MIME-Version: 1.0

The presence of this header field is an assertion that the message has been composed in compliance with the MIME specifications.

Since it is possible that in future the MIME specifications might extend the message format standard again, a formal BNF is given for the content of the MIME-Version field:

version := "MIME-Version" ":" 1*DIGIT "." 1*DIGIT

Thus, future format specifiers, which might replace or extend "1.0", are constrained to be two integer fields, separated by a period. If a message is received with a MIME-version value other than "1.0", it cannot be assumed to conform with RFC 2045.

Note that the MIME-Version header field is required at the top level of a message. It is not required for each body part of a multipart entity. It is required for the embedded headers of a body of type "message/rfc822" or "message/partial" if and only if the embedded message itself is claimed to be MIME-conformant.

It is not possible to fully specify how a mail reader that conforms with MIME as defined in RFC 2045 should treat a message that might arrive in the future with some value of MIME-Version other than "1.0".

It is also worth noting that version control for specific media types is not accomplished using the MIME-Version mechanism. In particular, some formats (such as application/postscript) have version-numbering conventions that are internal to the media format. Where such conventions exist, MIME does nothing to supersede them. Where no such conventions exist, a MIME media type might use a "version" parameter in the content-type field if necessary.

In the absence of a MIME-Version field, a receiving mail user agent (whether conforming to MIME requirements or not) may optionally choose to interpret the body of the message according to local conventions. Many such conventions are currently in use and it should be noted that in practice non-MIME messages can contain just about anything.

It is impossible to be sure that a non-MIME mail message is actually plain text in the US-ASCII character set since it might well be a message that, using some set of nonstandard local conventions that predate MIME, includes text in another character set or nontextual data presented in a manner that cannot be automatically recognised (e.g. a uuencoded compressed UNIX tar file).

3.5.3
Content-Type Header Field

The purpose of the Content-Type field is to describe the data contained in the body fully enough that the receiving UA can pick an appropriate agent or mechanism to present the data to the user, or otherwise deal with the data in an appropriate manner. The value in this field is called a media type.

The Content-Type header field specifies the nature of the data in the body of an entity by giving media type and subtype identifiers, and by providing auxiliary information that may be required for certain media types. After the media type and subtype names, the remainder of the header field is simply a set of parameters, specified in an

attribute=value

notation. The ordering of parameters is not significant.

In general, the top-level media type is used to declare the general type of data, while the subtype specifies a specific format for that type of data. Thus, a media type of "image/xyz" is enough to tell a user agent that the data is an image, even if the user agent has no knowledge of the specific image format "xyz". Such information can be used, for example, to decide whether or not to show a user the raw data from an unrecognised subtype—such an action might be reasonable for unrecognised subtypes of text, but not for unrecognised subtypes of image or audio. For this reason, registered subtypes of text, image, audio, and video should not contain embedded information that is really of a different type. Such compound formats should be represented using the "multipart" or "application" types.

Parameters are modifiers of the media subtype, and as such do not fundamentally affect the nature of the content. The set of meaningful parameters depends on the media type and subtype. Most parameters are associated with a single specific subtype. However, a given top-level media type may define parameters which are applicable to any subtype of that type. Parameters may be required by their defining content type or subtype or they may be optional.

An initial set of seven top-level media types is defined in RFC 2046. Five of these are discrete types whose content is essentially opaque as far as MIME processing is concerned. The remaining two are composite types whose contents require additional handling by MIME processors.

This set of top-level media types is intended to be substantially complete. It is expected that additions to the larger set of supported types can generally be accomplished by the creation of new subtypes of these initial types.

If another top-level type is to be used for any reason, it must be given a name starting with "X-" to indicate its nonstandard status and to avoid a potential conflict with a future official name.

The syntax of the content-type header field can be found in RFC 2045.

3.5.3.1
Content-Type Defaults

Default RFC 822 messages without a MIME Content-Type header considered to be plain text in the US-ASCII character set can be explicitly specified as

> Content-type: text/plain; charset=us-ascii

This default is assumed if no Content-Type header field is specified. It is also recommended that this default is assumed when a syntactically invalid Content-Type header field is encountered. In the presence of a MIME-Version header field and the absence of any Content-Type header field, a receiving UA can also assume that plain US-ASCII text was the sender's intent. Plain US-ASCII text may still be assumed in the absence of a MIME-Version or the presence of a syntactically invalid Content-Type header field, but the sender's intent might have been otherwise.

3.5.4
Content-Transfer-Encoding Header Field

Many media types which could be usefully transported via email are represented in their "natural" format, as 8bit character or binary data. Such data cannot be transmitted over some transfer protocols. For example, RFC 821 (SMTP) restricts mail messages to 7bit US-ASCII data with lines no longer than 1000 characters including any trailing CRLF line separator.

As a consequence it is necessary to define a standard mechanism for encoding such data into a 7bit short line format. Proper labelling of unencoded material in less restrictive formats for direct use over less restrictive transports is also desirable. Therefore RFC 2045 specifies the first time that such encodings will be indicated by a new "Content-Transfer-Encoding" header field and its syntax is specified in RFC 2045 as well.

3.5.4.1
Content-Transfer-Encodings Semantics

This single Content-Transfer-Encoding token actually provides two pieces of information. It specifies what sort of encoding transformation the body was subjected to and hence what decoding operation must be used to restore it to its original form, and it specifies what the domain of the result is.

The transformation part of any Content-Transfer-Encodings specifies, either explicitly or implicitly, a single, well-defined decoding algorithm, which for any sequence of encoded octets either transforms it to the original sequence of octets which was encoded, or shows that it is illegal as an encoded sequence. Content-Transfer-Encodings transformations never depend on any additional external profile information for proper operation. Note that while decoders must produce a single, well-defined output for a valid encoding, no such restrictions exist for encoders: Encoding a given sequence of octets to different, equivalently encoded sequences is perfectly legal.

Three transformations are currently defined: identity, the "quoted-printable" encoding, and the "base64" encoding. The domains are "binary", "8bit" and "7bit".

All the Content-Transfer-Encoding values "7bit", "8bit", and "binary" mean that the identity (i.e. NO) encoding transformation has been performed. As such, they serve simply as indicators of the domain of the body data, and provide useful information about the sort of encoding that might be needed for transmission in a given transport system.

The quoted-printable and base64 encodings transform their input from an arbitrary domain into material in the "7bit" range, thus making it safe to carry over restricted transports. The specific definition of the transformations are given below.

The proper Content-Transfer-Encoding label must always be used. Labelling unencoded data containing 8bit characters as "7bit" is not allowed, nor is labelling unencoded non-line-oriented data as anything other than "binary" allowed.

Unlike media subtypes, a proliferation of Content-Transfer-Encoding values is both undesirable and unnecessary. However, establishing only a single transformation into the "7bit" domain does not seem possible. There is a tradeoff between the desire for a compact and efficient encoding of largely binary data and the desire for a somewhat readable encoding of data that is mostly, but not entirely, 7bit. For this reason, at least two encoding mechanisms are necessary: a more or less readable encoding (quoted-printable) and a "dense" or "uniform" encoding (base64).

Mail transport for unencoded 8bit data is defined in RFC 1652. As of the initial publication of the MIME specification, there are no standardised Internet mail transports for which it is legitimate to include unencoded binary data in mail bodies. Thus there are no circumstances in which the "binary" Content-Transfer-Encoding is actually valid in Internet mail. However, in the event that binary mail transport becomes a reality in Internet mail, or when MIME is used in conjunction with any other binary-capable mail transport mechanism, binary bodies must be labelled as such using this mechanism.

3.5.5
Content-ID Header Field

In constructing a high-level user agent, it may be desirable to allow one body to make reference to another. Accordingly, bodies may be labelled using the "Content-ID" header field, which is syntactically identical to the "Message-ID" header field:

> id := "Content-ID" ":" msg-id

Like the Message-ID values, Content-ID values must be generated to be world-unique.

The Content-ID value may be used for uniquely identifying MIME entities in several contexts, particularly for caching data referenced by the message/external-body mechanism. Although the Content-ID header is generally optional, its use is mandatory in implementations which generate data of the optional MIME media type

> "message/external-body"

That is, each message/external-body entity must have a Content-ID field to permit caching of such data.

It is also worth noting that the Content-ID value has special semantics in the case of the multipart/alternative media type. This is explained in the section of RFC 2046 dealing with multipart/alternative.

3.5.6
Content-Description Header Field

The ability to associate some descriptive information with a given body is often desirable. For example, it may be useful to mark an "image" body as "a picture of the Space Shuttle Endeavor". Such text may be placed in the Content-Description header field. This header field is always optional.

> description := "Content-Description" ":" *text

The description is presumed to be given in the US-ASCII character set, although the mechanism specified in RFC 2047 may be used for non-US-ASCII Content-Description values as well.

3.5.7
Additional MIME Header Fields

Future documents may elect to define additional MIME header fields for various purposes. Any new header field that further describes the content of a message should begin with the string "Content-" to allow such fields which appear in a message header to be distinguished from ordinary RFC 822 message header fields.

> MIME-extension-field := <Any RFC 822 header field which begins with the
> string "Content-">

3.6
Post Office Protocol Version 3 (POP3)

3.6.1
Introduction

On certain types of smaller nodes in the Internet it is often impractical to maintain a MTS. For example, a workstation may not have sufficient resources (cycles, disk space) in order to permit a SMTP server and associated local mail delivery system to be kept resident and continuously running. Similarly, it may be expensive (or impossible) to keep a personal computer interconnected to an IP-style network for long amounts of time (the node lacks the resource known as "connectivity").

Despite this, it is often very useful to be able to manage mail on these smaller nodes, and they often support a UA to aid the tasks of mail handling. To solve this problem, a node which can support an MTS entity offers a maildrop service to these less endowed nodes. The Post Office Protocol—Version 3 (POP3), specified in RFC 1939, is intended to permit a workstation to dynamically access a maildrop on a server host in a useful fashion. Usually, this means that the POP3 protocol is used to allow a workstation to retrieve mail that the server is holding for it.

POP3 is not intended to provide extensive manipulation operations of mail on the server. The only thing it provides for is to download mail from the server to the client and to delete it afterwards at the server.

For the remainder of this chapter, the term "client host" refers to a host making use of the POP3 service, while the term "server host" refers to a host which offers the POP3 service.

3.6.2
Basic Operation

Initially, the server host starts the POP3 service by listening on TCP port 110. When a client host wishes to make use of the service, it establishes a TCP connection with the server host. When the connection is established, the POP3 server sends a greeting. The client and POP3 server then exchange commands and responses (respectively) until the connection is closed or aborted.

Commands in the POP3 consist of a case-insensitive keyword, possibly followed by one or more arguments. All commands are terminated by a CRLF pair. Keywords and arguments consist of printable ASCII characters. Keywords and arguments are each separated by a single SPACE character. Keywords are three or four characters long. Each argument may be up to 40 characters long.

Responses in the POP3 consist of a status indicator and a keyword possibly followed by additional information. All responses are terminated by a CRLF pair. Responses may be up to 512 characters long, including the terminating CRLF. There are currently two status indicators: positive ("+OK") and negative ("-ERR"). Servers must send the "+OK" and "-ERR" in upper case.

Responses to certain commands are multiline. In these cases, which are clearly indicated below, after sending the first line of the response and a CRLF, any additional lines are sent, each terminated by a CRLF pair. When all lines of the response have been sent, a final line is sent, consisting of a termination octet (decimal code 046, ".") and a CRLF pair. If any line of the multiline response begins with the termination octet, the line is "byte-stuffed" by pre-pending the termination octet to that line of the response. Hence a multiline response is terminated with the five octets "CRLF.CRLF". When examining a multiline response, the client checks to see if the line begins with the termination octet. If so and if octets other than CRLF follow, the first octet of the line (the termination octet) is stripped away. If so and if CRLF immediately follows the termination character, then the response from the POP server is ended and the line containing ".CRLF" is not considered part of the multiline response.

A POP3 session progresses through a number of states during its lifetime. Once the TCP connection has been opened and the POP3 server has sent the greeting, the session enters the AUTHORISATION state. In this state, the client must identify itself to the POP3 server. Once the client has successfully done this, the server acquires resources associated with the client's maildrop, and the session enters the TRANSACTION state. In this state, the client requests actions on the part of the POP3 server. When the client has issued the QUIT command, the session enters the UPDATE state. In this state, the POP3 server releases any resources acquired during the TRANSACTION state and says goodbye. The TCP connection is then closed.

A server must respond to an unrecognised, unimplemented, or syntactically invalid command with a negative status indicator. A server must respond to a command issued when the session is in an incorrect state with a negative status indicator as well. There is no general method for a client to distinguish between a server which does not implement an optional command and a server which is unwilling or unable to process the command.

A POP3 server may have an inactivity autologout timer. Such a timer must be of at least 10 minutes' duration. The receipt of any command from the client during that interval should suffice to reset the autologout timer. When the timer expires, the session does not enter the UPDATE state—the server should close the TCP connection without removing any messages or sending any response to the client.

3.6.3
The AUTHORISATION State

Once the TCP connection has been opened by a POP3 client, the POP3 server issues a one-line greeting. This can be any positive response. An example might be

S: +OK POP3 server ready

The POP3 session is now in the AUTHORISATION state. The client must now identify and authenticate itself to the POP3 server. Two possible mechanisms for doing this are described below, the USER and PASS command combination and the APOP command. Additional authentication mechanisms are described in RFC1734 "POP3 AUTHentication command". While there is no single authentication mechanism that is required by all POP3 servers, a POP3 server must of course support at least one authentication mechanism.

Once the POP3 server has determined through the use of any authentication command that the client should be given access to the appropriate maildrop, the POP3 server then acquires an exclusive-access lock on the maildrop which is necessary to prevent messages from being modified or removed before the session enters the UPDATE state. If the lock is successfully acquired, the POP3 server responds with a positive status indicator. The POP3 session now enters the TRANSACTION state, with no messages marked as deleted. If the maildrop cannot be opened for some reason (for example, a lock cannot be acquired, the client is denied access to the appropriate maildrop, or the maildrop cannot be parsed), the POP3 server responds with a negative status indicator. (If a lock was acquired but the POP3 server intends to respond with a negative status indicator, the POP3 server must release the lock prior to rejecting the command.) After returning a negative status indicator, the server may close the connection. If the server does not close the connection, the client may either issue a new authentication command and start again, or the client may issue the QUIT command.

After the POP3 server has opened the maildrop, it assigns a message-number to each message, and notes the size of each message in octets. The first message in the maildrop is assigned a message-number of "1", the second is assigned "2", and so on, so that the nth message in a maildrop is assigned a message-number of "n". In POP3 commands and responses, all message-numbers and message sizes are expressed in base-10 (i.e. decimal).

Here is the summary for the QUIT command when used in the AUTHORISATION state:

3.6.3.1
QUIT

Arguments: none

Restrictions: none

Possible Responses: +OK

3.6.4
The TRANSACTION State

Once the client has successfully identified itself to the POP3 server and the POP3 server has locked and opened the appropriate maildrop, the POP3 session is now in the TRANSACTION state. The client may now issue any of the following POP3 commands repeatedly. After each command, the POP3 server issues a response. Eventually, the client issues the QUIT command and the POP3 session enters the UPDATE state. Here are the POP3 commands valid in the TRANSACTION state:

- STAT
- LIST [msg]
- RETR msg
- DELE msg
- NOOP
- RSET

3.6.4.1
STAT

Arguments: none

Restrictions: may only be given in the TRANSACTION state

Discussion: The POP3 server issues a positive response with a line containing information for the maildrop. This line is called a "drop listing" for that maildrop.
In order to simplify parsing, all POP3 servers are required to use a certain format for drop listings. The positive response consists of "+OK" followed by a single space, the number of messages in the maildrop, a single space, and the size of the maildrop in octets. RFC 1939 makes no requirement on what follows the maildrop size. Minimal implementations should just end that line

of the response with a CRLF pair. More advanced implementations may include other information.

Note that RFC 1939 strongly discourages implementations from supplying additional information in the drop listing. Other, optional, facilities are discussed later on which permit the client to parse the messages in the maildrop. Additionally it is important to know that messages that are marked as deleted are not counted either in total.

Possible Responses: +OK nn mm

3.6.4.2
LIST [msg]

Arguments: a message-number (optional), which, if present, may not refer to a message marked as deleted

Restrictions: may only be given in the TRANSACTION state

Discussion: If an argument was given and the POP3 server issues a positive response with a line containing information for that message. This line is called a "scan listing" for that message.

If no argument was given and the POP3 server issues a positive response, then the response given is multi-line. After the initial +OK, for each message in the maildrop, the POP3 server responds with a line containing information for that message. This line is also called a "scan listing" for that message. If there are no messages in the maildrop, then the POP3 server responds with no scan listings--it issues a positive response followed by a line containing a termination octet and a CRLF pair.
In order to simplify parsing, all POP3 servers are required to use a certain format for scan listings. A scan listing consists of the message-number of the message, followed by a single space and the exact size of the message in octets. Methods for calculating the exact size of the message are described in the "Message Format" section below. RFC 1939 does not specify any requirement on what follows the message size in the scan listing. Minimal implementations should just end that line of the response with a CRLF pair. More advanced implementations may include other information, as parsed from the message.
Note that RFC 1939 strongly discourages implementations from supplying additional information in the scan listing. Other,

optional, facilities are discussed later on which permit the client to parse the messages in the maildrop. Additionally it is important to know that messages marked as deleted are not listed.

Possible Responses: +OK scan listing follows
                          -ERR no such message

3.6.4.3
RETR msg

Arguments: a message-number (required) which may not refer to a message marked as deleted

Restrictions: may only be given in the TRANSACTION state

Discussion: If the POP3 server issues a positive response, then the response given is multiline. After the initial +OK, the POP3 server sends the message corresponding to the given message-number, being careful to byte-stuff the termination character (as with all multiline responses).

Possible Responses: +OK message follows
                          -ERR no such message

3.6.4.4
DELE msg

Arguments: a message-number (required) which may not refer to a message marked as deleted

Restrictions: may only be given in the TRANSACTION state

Discussion: The POP3 server marks the message as deleted. Any future reference to the message-number associated with the message in a POP3 command generates an error. The POP3 server does not actually delete the message until the POP3 session enters the UPDATE state.

Possible Responses: +OK message deleted
                          -ERR no such message

3.6.4.5
NOOP

Arguments: none

Restrictions: may only be given in the TRANSACTION state

Discussion: The POP3 server does nothing, it merely replies with a positive
response.

Possible Responses: +OK

3.6.4.6
RSET

Arguments: none

Restrictions: may only be given in the TRANSACTION state

Discussion: If any messages have been marked as deleted by the POP3 server,
they are unmarked. The POP3 server then replies with a positive
response.

Possible Responses: +OK

3.6.5
The UPDATE State

When the client issues the QUIT command from the TRANSACTION state, the
POP3 session enters the UPDATE state. (Note that if the client issues the QUIT
command from the AUTHORISATION state, the POP3 session terminates but
does not enter the UPDATE state.)

If a session terminates for some reason other than a client-issued QUIT
command, the POP3 session does not enter the UPDATE state and must not
remove any messages from the maildrop.

3.6.5.1
QUIT

Arguments: none

Restrictions: none

Discussion: The POP3 server removes all messages marked as deleted from the maildrop and replies as to the status of this operation. If there is an error, such as a resource shortage, encountered while removing messages, the maildrop may result in having some or none of the messages marked as deleted be removed. In no case may the server remove any messages not marked as deleted.

Whether the removal was successful or not, the server then releases any exclusive-access lock on the maildrop and closes the TCP connection.

Possible Responses: +OK
                         -ERR some deleted messages not removed

3.6.6
Optional POP3 Commands

The POP3 commands discussed above must be supported by all minimal implementations of POP3 servers.

The optional POP3 commands described below permit a POP3 client greater freedom in message handling, while preserving a simple POP3 server implementation.

NOTE: RFC 1939 strongly encourages implementations to support these commands in lieu of developing augmented drop and scan listings. In short, the philosophy of RFC 1939 is to put intelligence in the part of the POP3 client and not the POP3 server.

3.6.6.1
TOP msg n

Arguments: a message-number (required) which may not refer to to a message marked as deleted, and a non-negative number of lines (required)

Restrictions: may only be given in the TRANSACTION state

Discussion: If the POP3 server issues a positive response, then the response given is multiline. After the initial +OK, the POP3 server sends the headers of the message, the blank line separating the headers from the body, and then the number of lines of the indicated message's body, being careful to byte-stuff the termination character (as with all multiline responses).

Note that if the number of lines requested by the POP3 client is greater than the number of lines in the body, then the POP3 server sends the entire message.

Possible Responses: +OK top of message follows
                         -ERR no such message

3.6.6.2
UIDL [msg]

Arguments: a message-number (optional), which, if present, may not refer to a message marked as deleted

Restrictions: may only be given in the TRANSACTION state.

Discussion: If an argument was given and the POP3 server issues a positive response with a line containing information for that message, this line is called a "unique-id listing" for that message.
If no argument was given and the POP3 server issues a positive response, then the response given is multiline. After the initial +OK, for each message in the maildrop, the POP3 server responds with a line containing information for that message. This line is called a "unique-id listing" for that message.
In order to simplify parsing, all POP3 servers are required to use a certain format for unique-id listings. A unique-id listing consists of the message-number of the message, followed by a single space and the unique-id of the message. No information follows the unique-id in the unique-id listing.
The unique-id of a message is an arbitrary server-determined string, consisting of one to 70 characters in the range 0x21 to 0x7E, which uniquely identifies a message within a maildrop and which persists across sessions. This persistence is required even if a session ends without entering the UPDATE state. The server should never reuse an unique-id in a given maildrop, for as long as the entity using the unique-id exists.
Note that messages marked as deleted are not listed.
While it is generally preferable for server implementations to store arbitrarily assigned unique-ids in the maildrop, RFC 1939 is intended to permit unique-ids to be calculated as a hash of the message. Clients should be able to handle a situation where two identical copies of a message in a maildrop have the same unique-id.

Possible Responses: +OK unique-id listing follows

-ERR no such message

3.6.6.3
USER Name

Arguments: a string identifying a mailbox (required), which is of significance only to the server

Restrictions: may only be given in the AUTHORISATION state after the POP3 greeting or after an unsuccessful USER or PASS command

Discussion: To authenticate using the USER and PASS command combination, the client must first issue the USER command. If the POP3 server responds with a positive status indicator ("+OK"), then the client may issue either the PASS command to complete the authentication, or the QUIT command to terminate the POP3 session. If the POP3 server responds with a negative status indicator ("-ERR") to the USER command, then the client may either issue a new authentication command or may issue the QUIT command.

The server may return a positive response even though no such mailbox exists. The server may return a negative response if the specified mailbox exists, but does not permit plain text password authentication.

Possible Responses: +OK name is a valid mailbox
                         -ERR never heard of mailbox name

3.6.6.4
PASS String

Arguments: a server/mailbox-specific password (required)

Restrictions: may only be given in the AUTHORISATION state immediately after a successful USER command

Discussion: When the client issues the PASS command, the POP3 server uses the argument pair from the USER and PASS commands to determine if the client should be given access to the appropriate maildrop.

Since the PASS command has exactly one argument, a POP3 server may treat spaces in the argument as part of the password, instead of as argument separators.

Possible Responses: +OK maildrop locked and ready
                         -ERR invalid password
                         -ERR unable to lock maildrop

3.6.6.5
APOP Name D

Arguments: a string identifying a mailbox and a MD5 digest string (both required)

Restrictions: may only be given in the AUTHORISATION state after the POP3 greeting or after an unsuccessful USER or PASS command

Discussion: Normally, each POP3 session starts with a USER/PASS exchange. This results in a server/user-id-specific password being sent in the clear on the network. For intermittent use of POP3, this may not introduce a sizeable risk. However, many POP3 client implementations connect to the POP3 server on a regular basis— to check for new mail. Further, the interval of session initiation may be on the order of five minutes. Hence, the risk of password capture is greatly enhanced.

An alternate method of authentication is required which provides for both origin authentication and replay protection, but which does not involve sending a password in the clear over the network. The APOP command provides this functionality.

A POP3 server which implements the APOP command will include a timestamp in its banner greeting. The syntax of the timestamp corresponds to the 'msg-id' in RFC 822, and must be different each time the POP3 server issues a banner greeting. For example, on a UNIX implementation in which a separate UNIX process is used for each instance of a POP3 server, the syntax of the timestamp might be

<process-ID.clock@hostname>

where 'process-ID' is the decimal value of the process' PID, clock is the decimal value of the system clock, and hostname is the fully qualified domain-name corresponding to the host where the POP3 server is running.

The POP3 client makes note of this timestamp, and then issues the APOP command. The 'name' parameter has identical semantics to the 'name' parameter of the USER command. The 'digest' parameter is calculated by applying the MD5 algorithm specified in RFC1321 to a string consisting of the timestamp (including angle brackets) followed by a shared secret. This shared secret is a string known only to the POP3 client and server. Great care should be taken to prevent unauthorised disclosure of the secret, as knowledge of the secret will allow any entity to successfully masquerade as the named user. The 'digest' parameter itself is a 16-octet value which is sent in hexadecimal format, using lower-case ASCII characters.

When the POP3 server receives the APOP command, it verifies the digest provided. If the digest is correct, the POP3 server issues a positive response, and the POP3 session enters the TRANSACTION state. Otherwise, a negative response is issued and the POP3 session remains in the AUTHORISATION state.

Note that as the length of the shared secret increases, so does the difficulty of deriving it. As such, shared secrets should be long strings.

Possible Responses: +OK maildrop locked and ready
                         -ERR permission denied

3.6.7
Scaling and Operational Considerations

Since some of the optional features described above were added to the POP3 protocol, experience has accumulated in using them in large-scale commercial post office operations where most of the users are unrelated to each other. In these situations and others, users and vendors of POP3 clients have discovered that the combination of using the UIDL command and not issuing the DELE command can provide a weak version of the "maildrop as semipermanent repository" functionality normally associated with IMAP. Of course the other capabilities of IMAP, such as polling an existing connection for newly arrived messages and supporting multiple folders on the server, are not present in POP3.

When these facilities are used in this way by casual users, there is a tendency for already-read messages to accumulate on the server without bound. This is clearly an undesirable behaviour pattern from the standpoint of the server operator. This situation is aggravated by the fact that the limited capabilities of the POP3 do not permit efficient handling of maildrops having hundreds or thousands of messages. Consequently, it is recommended that operators of large-scale multiuser servers, especially those in which the user's only access to the maildrop is via POP3, consider such options as

• imposing a per-user maildrop storage quota or the like.

A disadvantage to this option is that the accumulation of messages may result in the user's inability to receive new ones into the maildrop. Sites which choose this option should be sure to inform users of impending or current exhaustion of quota, perhaps by inserting an appropriate message into the user's maildrop.

- enforce a site policy regarding mail retention on the server.

Sites are free to establish local policy regarding the storage and retention of messages on the server, both read and unread. For example, a site might delete unread messages from the server after 60 days and delete read messages after 7 days. Such message deletions are outside the scope of the POP3 protocol and are not considered a protocol violation.

Server operators enforcing message deletion policies should take care to make all users aware of the policies in force.

Clients must not assume that a site policy will automate message deletions, and should continue to explicitly delete messages using the DELE command when appropriate.

It should be noted that enforcing site message deletion policies may be confusing to the user community, since their POP3 client may contain configuration options to leave mail on the server which will not in fact be supported by the server.

One special case of a site policy is that messages may only be downloaded once from the server, and are deleted after this has been accomplished. This could be implemented in POP3 server software by the following mechanism: "following a POP3 login by a client which was ended by a QUIT, delete all messages downloaded during the session with the RETR command". It is important not to delete messages in the event of abnormal connection termination (i.e. if no QUIT was received from the client) because the client may not have successfully received or stored the messages. Servers implementing a download-and-delete policy may also wish to disable or limit the optional TOP command, since it could be used as an alternate mechanism to download entire messages.

3.6.8
POP3 Command Summary

3.6.8.1
Minimal POP3 Commands

- USER name   valid in the AUTHORISATION state
- PASS string
- QUIT
- STAT   valid in the TRANSACTION state

- LIST [msg]
- RETR msg
- DELE msg
- NOOP
- RSET
- QUIT

3.6.8.2
Optional POP3 Commands

- APOP name digest   valid in the AUTHORISATION state
- TOP msg n   valid in the TRANSACTION state
- UIDL [msg]

3.6.8.3
POP3 Replies

+OK
-ERR

Note that with the exception of the STAT, LIST, and UIDL commands, the reply given by the POP3 server to any command is significant only to "+OK" and "-ERR". Any text occurring after this reply may be ignored by the client.

3.6.8.4
Example POP3 Session

```
S: <wait for connection on TCP port 110>
C: <open connection>
S: +OK POP3 server ready <1896.697170952@dbc.mtview.ca.us>
C: APOP mrose c4c9334bac560ecc979e58001b3e22fb
S: +OK mrose's maildrop has 2 messages (320 octets)
C: STAT
S: +OK 2 320
C: LIST
S: +OK 2 messages (320 octets)
S: 1 120
S: 2 200
S: .
C: RETR 1
S: +OK 120 octets
S: <the POP3 server sends message 1>
S: .
C: DELE 1
```

S: +OK message 1 deleted
C: RETR 2
S: +OK 200 octets
S: <the POP3 server sends message 2>
S: .
C: DELE 2
S: +OK message 2 deleted
C: QUIT
S: +OK dewey POP3 server signing off (maildrop empty)
C: <close connection>
S: <wait for next connection>

3.6.8.5
Message Format

All messages transmitted during a POP3 session are assumed to conform to the standard for the format of Internet text messages specified in RFC822.

It is important to note that the octet count for a message on the server host may differ from the octet count assigned to that message due to local conventions for designating end-of-line. Usually, during the AUTHORISATION state of the POP3 session, the POP3 server can calculate the size of each message in octets when it opens the maildrop. For example, if the POP3 server host internally represents end-of-line as a single character, then the POP3 server simply counts each occurrence of this character in a message as two octets. Note that lines in the message which start with the termination octet need not (and must not) be counted twice, since the POP3 client will remove all byte-stuffed termination characters when it receives a multiline response.

4 X.400—Internet Mail Gateways

4.1
Introduction

One of the most important features required in an X.400 messaging environment is the provision of Internet gateways, in order to make communication between the huge number of Internet users and X.400 users possible. The large community of people using RFC 822 based mail services have to be considered as well in order to get a critical mass of users.

Because of the growing demand for the best and smoothest possible interworking between X.400 based and SMTP/MIME based users, not only the IETF specifications have to be changed in order to correspond to this situation but the same efforts also have to be made on the X.400 side. There are some aspects of the interworking which might be improved by changing ITU-T Recommendation X.400 (ISO/IEC 1021). This fact has been realised within the ITU/ISO so that there is a new work item out for ballot covering at least the following topics:

- exchange of secured data between end-users located on both messaging systems;
- specification of Internet-style addresses in O/R addresses;
- use of Internet-defined document types (e.g. HTML) in X.400 IPM;
- mapping between X.400 body part types and MIME content-types;
- support of emerging SMTP and MIME facilities.

This work item will also lead to the production of a technical report listing all the specifications available in ITU-T, ISO, and IETF, which are relevant when considering the problem of interworking between X.400 and SMTP/MIME: This technical report will also aim at providing a tutorial text on those various specifications, regarding what their scope is, which aspects each of them covers, and which problems they might bring.

In order to give an overview of the current status, latest efforts done within the IETF, especially within the MIME X.400 Enhanced Relay (MIXER) working group, were collected as well. Furthermore, an analysis of current X.400 products was carried out, with respect to their internet gateway features. Finally the current status of ADMD—Internet connections within Europe was analysed. Results are presented in the following sections.

Although RFC 822 is the basis of several message transfer protocols, the Simple Mail Transfer Protocol (SMTP) is of particular importance. Throughout this section "gateway" stands for "SMTP/MIME - X.400 gateway".

Three categories of functions are usually performed by gateways, which are illustrated in the following figure:

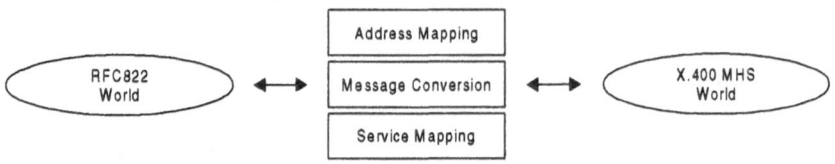

Fig. 4.1. Gateway functions

Address Mapping is considered to be the most complicated part of the gateway functionality. Since both sides of the gateway have their own address structures, parts which cannot be mapped at all are still under study.

Message Conversion takes care of the fact that some body parts at X.400 side do not have their pendants at the SMTP side and vice versa.

Service Mapping describes how to support services like delivery notifications or receipt notifications, which are well known in the X.400 world, at the other side of the fence.

Several RFCs were published on X.400 - Internet gateways, the most important of which were RFC 1327 and its companion 1495. In recent times an updated version has been introduced by an IETF Working Group: MIXER (MIME Internet X.400 Enhance Relay). The specifications produced by this Working Group have been adopted by almost all vendors and the core specification of MIXER is an accepted way of performing mapping between X.400 and Internet mail.

Because of its importance, the next section is dedicated to MIXER and will summarise its most important features.

4.2
MIME Internet X.400 Enhanced Relay (MIXER)

As mentioned above the MIXER specifications are widely accepted and adopted by the industry. All of the keyplayers in the X.400 filed have already or plan to implement at least parts of MIXER. Additionally they are all implementing the new address mapping concept MCGAM, introduced in the next section.

MIXER describes a set of mappings which will enable interworking between systems operating the CCITT (ITU-T) X.400 Series of Recommendations on Message Handling Systems (1984, 1988, 1992 and 1996 versions)/ISO IEC 10021 Message Oriented Text Interchange Systems (MOTIS), and systems using the SMTP/MIME protocol and RFC 822 message formats.

It should be noticed that MIXER works with every transfer protocol based on RFC822. However, as it is of particular importance, main emphasis is put on SMTP (RFC 821).

4.2.1
Main Features

The main features of MIXER implementations are the following:

- Refers to Interpersonal Messaging; human-end-users are assumed. MIXER is not defined for other message types, e.g. EDI;
- IPNs (Interpersonal Notifications), the X.400 delivery notifications are handled on the SMTP/MIME side as well. They are mapped on the MIME message;
- A new way of handling mapping rules is introduced: MCGAM (MIXER Conformant Global Address Mapping);
- Most header elements have a natural mapping between both worlds;
- Some X.400 header elements that do not have their pendants at the MIME header are mapped on the MIME message;
- Most mapping is transparent to the user;
- MIXER provides a framework of how two applications on both sides, X.400 and SMTP/MIME, could work together.

4.2.1.1
X.400 Features That Cannot be Mapped

The following X.400 features cannot be mapped to SMTP:

- EDI, according to X.435;
- Security features, because they are very protocol dependent;
- X.400 probes are not known in MIME, however, they are supported at the gateway, rather than being transferred into the MIME world.

4.2.1.2
MIXER Conformant Global Address Mapping (MCGAM)

Address mapping is often considered to be the most complicated part of mapping between X.400 and SMTP, thus, address-mapping technology is crucial to gateways.

Information is needed by the gateway how to map RFC 822 addresses onto X.400 O/R addresses and vice versa. Therefore mapping rules are collected in mapping tables, which build the information basis for the gateway. Mapping by means of such tables only work well if all gateways in the world use the same tables.

Till now, the updating of the mapping tables has been carried out in a rather centralised way. On the global level, the collection and distribution of RFC 1327 address mapping tables has been coordinated by the MHS Coordination Service, situated in Switzerland.

There have been mainly three kinds of tables:

- Mapping Table 1 (map1, table1 X2R) which consisted of rules that carry out X.400 -> RFC 822 mapping;
- Mapping Table 2 (map2, table2, R2X) which consisted of rules that carry out RFC 822 -> X.400 mapping;
- Gateway table (GW) which is scanned, if no mapping table2 entry is found.

One could, of course, imagine that this procedure caused a lot of administrative work and serious problems.

MIXER removed the requirement to use or maintain a set of global centralised mapping. Mapping information can now be managed in a distributed rather than in a centralised way. Organisations can publish their MIXER mapping or preferred gateway routing information using just local resources (their local DNS server), avoiding the need for a strong coordination with any centralised organisation. MIXER conformant gateways and tools located on Internet hosts can retrieve the mapping information querying the DNS instead of having fixed tables which need to be centrally updated and distributed.

Another possibility to publish MIXER mapping information will be the use of X.500 directory systems, which is of particular interest for any MHS X.400 system. An X.500 directory system has to be implemented anyway in order to provide a comfortable X.400 MHS Mail System.

4.2.1.3
MIME Body Parts

The following list gives an overview of currently used MIME body parts:

- Text/plain
- Text/html
- Image/gif
- Image/jpeg
- Image/tiff
- Application/activemessage
- Application/edifact
- Application/edi-X12

- Application/edit-consent
- Application/mac-bihex40
- Application/msword
- Application/postscript
- Application/rtf
- Application/zip
- Video/mpeg
- Video/quicktime
- Video/x-msvideo
- Audio/basic

4.2.1.4
Conversion Tables

The following tables list currently defined conversions between X.400 and RFC 822 body parts. Basically there are 4 types of conversion

Table 1. Conversion types

Conversion Types	Explanation
Byte copy	No conversion is performed. The byte stream is simply copied between X.400 and MIME.
Text conversion	The text is scanned and reformatted. This conversion is carried out when mapping by simply copying byte by byte is not possible.
Image conversion	Is used for conversion of e.g. G3 facsimiles or JPEG byte streams. Again the byte stream is scanned and reformatted.
Tunnelling	No conversion is carried out. Instead, an encapsulation is performed.

4.2.1.5
X.400—MIME Conversion Table

In order to perform conversion, two tables are needed, one for each direction (asymmetrical conversion).

Table 2. X.400-MIME Body part conversion

X.400 body part	MIME body part
Ia5Text	Text/plain; US-ASCII or ISO 8859-x characterset
Voice	No mapping
G3-facsimile	G3fax
G4 class1	No mapping
teletex	Text/plain; characterset= teletex
videotex	No mapping
encrypted	No mapping
Bilaterally defined	Application/octet stream
Nationally defined	No mapping
Forwarded IPM	Message/RFC822 or multipart
General text	Text/plain; characterset = ISO 8859-x
Mime-postscript-body	Application/postscript
Mime-jpeg-body	Image/jpeg
Mime-gif-body	Image/gif
FTAM	Various
ODA	Application/oda

4.2.1.6
MIME—X.400 conversion table

Table 3. MIME—X.400 Body part conversion

MIME body part	X.400 body part
Text/plain; US-ASCII characterset	IA5Text
Text/plain; ISO 8859-x characterset	EBP-General text
Text/richtext	No mapping
Application/oda	EBP - oda
Applications/octet-streams	Bilaterally defined or FTBP unknown attachment
Application/postscript	EBP - MIME-postscript-body
Image/G3fax	G3-facsimile
Image/jpeg	EBP - mime-jpeg-body-partImage/gif
	EBP - mime gif-body-part
Audio/basic	No mapping
Video/mpeg	No mapping
Message RFC822	Forwarded IPM
multipart	ForwardedIPM
Multipart/encrypted	Encapsulation
Multipart/signed	encapsulation

5 Directory Services

Directories play a significant role in Open Systems Interconnection, whose aim is to allow, with a minimum of technical agreement outside of the interconnection standards themselves, the interconnection of information processing systems:

- from different vendors;
- under different managements;
- of different levels of complexity;
- of different ages.

Directories can be used for a wide range of applications if two computer systems want to communicate with each other. Their beginning was in the 1980 when the first email systems were developed. This close relationship between directories and email services results from the fact that directories can be used for storing email addresses in a user friendly manner. But storing the email addresses in a more user friendly manner is not the only kind of usage for directories in mail environments. Another one would be to store the properties of the users of the service in order to know which file formats and attachments the user can handle.

But directories can be used for other applications as well. In recent times electronic commerce over the Internet is a very heavily discussed topic. Because the data sent are sometimes important and need to be kept private, security technologies have to be applied. The most often used technologies in this area are digital signature and encryption. Directories can be used to store the public keys of the users of a service to make them available as widely and easily as possible.

There are several different directory specifications available. The two most important and most widely used ones are the X.500 Series of Recommendations, published by the ITU-T, and the Lightweight Directory Access Protocol(LDAP), published by the Internet Engineering Task Force.

5.1
The X.500 Series of Recommendations

The need for X.500 directories was determined during the definition of the 1984 X.400 standards. The primary impetus behind the definition of X.500 comes from a desire of to provide better support for naming and addressing in large distributed X.400 message handling systems.

Since the use of directory services was introduced to X.400 in MHS 1988 as an optional feature, alternatives have been defined in the standard in order to handle most important functions even without an X.500 service on the network. This is true, for example, for distributed lists (DLs), which may operate both in conjunction with, and in the absence of, X.500 directory services. In general, however, when X.500 directory services are present, functions are handled more elegantly and transparently from the user's point of view.

Keeping this in mind, almost all vendors of X.400 products have implemented an X.500 directory within their product portfolio. Similar to X.400, the initial version of X.500, which was published in 1988, had some significant drawbacks. In 1993 a new version which had potential changes and improvements related to the initial version of 1988 was published.

5.1.1
The X.500 Series of Recommendations 1988

The X.500 Series of Recommendations 1988 consists of the following documents:

- X.500 The Directory: Overview of Concepts, Models and Services
- X.501 The Directory: Models
- X.509 The Directory: Authentication Framework
- X.511 The Directory: Abstract Service Definition
- X.518 The Directory: Procedures for Distributed Operations
- X.519 The Directory: Protocol Specifications
- X.520 The Directory: Selected Attribute Types
- X.521 The Directory: Selected Object Classes
- X.525 The Directory: Replication

5.1.2
The X.500 Series of Recommendations 1993

The X.500 Series of Recommendations 1993 consists of the following documents:

- X.500 The Directory: Overview of Concepts, Models and Services
- X.501 The Directory: Models
- X.509 The Directory: Authentication Framework
- X.511 The Directory: Abstract Service Definition
- X.518 The Directory: Procedures for Distributed Operations
- X.519 The Directory: Protocol Specifications
- X.520 The Directory: Selected Attribute Types

- X.521 The Directory: Selected Object Classes
- X.525 The Directory: Replication

5.1.3
X.500 Functional Model

The X.500 Functional Model consists of three principal functional components:

- The directory information base (DIB);
- directory system agents (DSAs);
- directory user agents (DUAs).

The DIB contains the collection of information about users, resources, and the network that is maintained by the directory. The DIB resides physically within and is managed by network servers called DSAs. DSAs provide the actual directory service and implement the server side of the directory operations. DUAs represent the "client" side of the directory service. They represent the user in accessing the information stored in the directory. As the information contained in the directory grows, it is usually necessary to partition the DIB amongst multiple DSAs, called cooperating DSAs. This is done to increase availability of the information and improve overall system performance by ensuring that information is maintained close to the network users who need access to it most often. From the perspective of the DUAs, however, such a collection of cooperating DSAs must continue to behave as a single unified database. In other words, a directory query addressed to any DSA that is part of the directory service environment must yield the same results as the same query to any other DSA in that same environment.

In the initial version of X.500 published in 1988 only two protocols were defined:

- The directory access protocol (DAP), which is used by DUAs to access information stored in DSAs;
- the directory system protocol (DSP), which is used between DSAs to service user queries that require information that might be distributed over multiple DSAs.

In the X.500 1993 two additional protocols were defined for communication between DSAs:

- The directory information shadowing protocol (DISP), which defines how information and updates are shared between DSAs;
- The directory operational binding management protocol (DOP), which enables DSAs to negotiate the nature of binding agreements, and parameters like frequency of updates.

Since usage of the latter two protocols was introduced to X.500 for optional use, they are not used by all of the implementations. Especially the DOP Protocol is implemented very seldom.

5.2
Lightweight Directory Access Protocol (LDAP)

LDAP arose from the first experiences when implementing X.500 services on the Internet from 1989 to 1991. The first LDAP standard, known as LDAP v1 was developed in 1993 and published as RFC 1487, which was later updated to LDAP v2 in 1995 and published as RFC 1777. The most important and most widely used LDAP implementation is certainly the freely available one of the University of Michigan.

The intention of LDAP was to access the X.500 directory in an easier way than the X.500 DAP protocol does. Easier in this relation means

*   Simple protocol encoding
*   Names and attributes use text encoding
*   LDAP is mapped directly over TCP/IP
*   Availability of a simpler API
*   LDAP relies on X.500 for the service definition and distributed objects

The reason for the significantly growing interest in LDAP came from the announcement of Netscape and many other vendors of directory systems that they will adopt LDAP in their products. Because of the fact that LDAP v2 does not have some very important features of X.500 (93), in 1996 a new version of LDAP, LDAP v3, was developed and discussed by the Access, Searching and Indexing (ASID) Working Group of the IETF. The new features of LDAP v3 are

*   Access to X.500 (93) features
*   Improved security features because of the use of SSL
*   Extensibility mechanisms
*   Referrals

Another new feature of LDAP v3 is that X.500 is no longer needed as the back end making it possible to use other mainly proprietary directory servers. This change is sometimes misunderstood in the sense that people have the opinion that LDAP has now lost the complexity and heaviness of X.500 baggage. This is not true because most of the LDAP specification relies on X.500, especially the service definitions as mentioned above.

For the future LDAP will certainly be the most important protocol for accessing directory systems. But X.500 will be most appropriate on the server side and especially for Intranet directories.

6 Electronic Commerce and Electronic Data Interchange

What is Electronic Commerce (EC)? Electronic Commerce is defined as an integrated arrangement of business practices and processes, technical application configurations, and organisational structures that utilise Electronic Data Interchange.

EC began in 1845 with the first commercial operation of the telegraph which used coding for characters. Yet it is the significant availability of integrated similar applications, infrastructure and technologies that has made the term EC popular today. EC seems to be the best fit to describe a wide range of interrelated applications and technologies. The prerequisites for EC are the following:

- It must be conducted via electronic communications.
- It must include buyers.
- It must include sellers.
- It must include a delivery mechanism.
- It must include payment.

Before describing some of the trends and technologies forming modern day EC, it is worth going back to the 19th Century to review the intrinsic market dynamics associated with the improvements to the process known as commerce, all of which apply today and lead to the following visions:

- Support electronic commerce with millions of potential private, commercial, and government trading partners (TPs).
- Communicate with TPs through commercial value-added networks or through the Internet.
- Provide the ability to sign, seal, secure, and postmark TP transactions electronically.
- Incorporate open industry standards for TP communications and security.
- Maintain a secure archive and audit trail of TP communications.

6.1
The History of Electronic Commerce

We look back to the middle of the 19th Century where the introduction of the telegraph in 1845 gave buyers and sellers the first opportunity to conduct commerce without being physically co-located. The invention of the telegraph allowed companies and individuals to conduct commerce without having to physically meet. The telegraph also allowed information to be exchanged in a faster way than any other mechanism in existence at the time. This information was not restricted to generic information, but also to specific supporting applications, like money transfer. The ability of buyers and sellers not having to be co-located to conduct commerce was, and is today, a major driver in EC. Equally, technologies possessing the ability to support a faster process for commerce has always been, and is still today, a major driver in EC. For both, the buyer and the seller, the benefits of EC were and still are the following:

- a wider choice;
- a wider market.

Additionally, it must not be forgotten that EC was being enabled not only between anonymous companies and people, but also between buyers and sellers of different divisions of the same company. Keeping this in mind there are five "need rules" that form the basis for the user requirements and expectations of EC:

- EC supports buyer and seller geographical independence.
- EC enables commerce to operate faster.
- EC applies both inter- and intra-enterprise applications.

  The advent of the telephone in 1875 further increased the speed at which EC could be undertaken. Not only did the telephone allow the buyer and seller to conduct commerce, it also allowed them to query the transaction as part of the same interactive session. The need for interactive functionality remains today, particularly for EC with the "common man". However, even the more sophisticated EDI applications still need to rely on interactive voice.

- EC should also support interactive needs.

  The use of the telegraph and the telephone were the only forms of EC until the advent of commercial computing in the early 1950s. The computer led to a new mechanism for enabling a more efficient process in the previously administrative-back-office-side of EC. The major applications were in the automated ticketing and supply-chain management areas. Throughout the 1950s and 1960s, these systems were mainframe-based and connected to the users via direct host links, or for "people users" through teletype operation. Such systems did not go directly to the buying public but through market intermediaries. The primary benefits of these EC systems were enhancing not only the timeliness of EC but also reducing the administrative costs of supporting commerce. The use of computerisation in this mode provided the first instances of business process re-engineering (BPR)—acknowledging that the paper-driven process of the day could be displaced by more efficient methods.

- EC allows business process re-engineering to improve efficiency.

By the late 1970s, additional forms of EC were being implemented—formal EDI (the full computer application-to-application automated transfer of data), facsimile (an extension of the telephone), and electronic mail. During this time, the rise of the value-added network (VAN) for supporting business-to-business applications was also seen. Initially VANs were established as a result of specific sector needs, for example, the automotive industry. They soon evolved to support generic EDI and derivative applications for trading between different companies. Indeed, VANs played an important role in enabling EC, since they provided the specialist applications and communications needed by a trading community. The primary focus of many VAN applications has always been and still is the enhancement of the supply chain process. By the late 1980s, VANs were the dominant mechanism for supporting EC between major trading companies, their divisions and their partners across a range of technologies. Where EC was implemented for use by the "common man", it was mainly restricted to telephone- based services—the telephone and the facsimile. Witness the growth of telephone banking, facsimile bureaus and other similar technologies. Early trials of teleprinter-to-bureau and PC-based EC were not successful, and even government-sponsored videotex projects soon reached limits of integration and saturation. The primary reason for the lack of sophisticated EC at the "common man" level was a combination of high networking costs to join trading communities, difficulty in joining other communities, and the generally low penetration of computing power. By the late 1980s, EC could be said to have polarised into two classes of users:

- Business-to-business via EDI, e-mail and telephone.
- Business-to-consumer via facsimile and telephone.

This split in technologies and users became quite entrenched, and it seemed this picture would remain. The split led to a natural tension between the two pools of users with the business-to-business processes, supported by EDI and e-mail being unable to extend to smaller companies and users. Indeed, main business-to-business EDI/EC systems were unable to even penetrate the small-to-medium-sized enterprises (SMEs) sector. The few hybrid solutions that evolved, for example, EDI-to-fax, did not improve the situation by much.

6.2
Electronic Commerce Today

Today, EC is facing the most exciting and interesting phase since its beginning more than 150 years ago. Not only does it incorporate a wealth of interrelated applications—facsimile, electronic mail, EDI and so on, it is also experiencing a massive expansion in the number of people and companies that can make use of the developing infrastructure. Until a few years ago, EC was restricted to the relatively few major trading companies who used EDI for improving the supply-chain process. Despite articles and papers galore extolling the virtues of using EDI, in addition to the work of governments in this area, the total number of companies achieving EDI remains small, and the number of ordinary people undertaking EC is equally small and focused on voice and facsimile. What has changed now? The answer is the Internet. Since the Internet is available in more that 160 countries all over the world, that access from any two points of the globe is relatively straight-forward. More than that, the support for the Internet from major service providers is such that the issue of cost is not a primary consideration. Few, if any, Internet service providers are profiting on their operations since they are engaged in the strategic buying of market share. As a result, users can gain access to the Internet for relatively low costs, and certainly lower than by using traditional VANs. For the serious business user, the Internet today does not provide the quality of service needed for business-to-business EC. However, it does allow for EC between large organisations, smaller businesses and consumers where these requirements are less stringent. Serious business users still require guaranteed service levels and known trading partners due to their automated processes being sufficiently refined to support in-depth application-to-application inter-working. For inter-company EC, which is today mainly served by traditional EDI, VANs will continue to offer industrial strength services to major hub-oriented users. These VANs will also offer such customers front-office services based on the Web and Internet that build a seamless link to existing EDI processes. Such VANs will also offer an increasing integration of other EC supporting applications to provide a complete portfolio of services. The more enterprising VANs will also involve themselves in customer site services that will enable both back-office and front-office services for users, and some of these services may well bypass the VAN network itself as the march to a disintermediated market continues. As such, VANs will also provide the new types of value-added service as needed in this market and these services will focus more on supporting security and financial payment systems than on network connectivity and message conversion.

6.3 Electronic Data Interchange—An Introduction

There has been a lot of talk about EDI in recent times. So what does EDI mean and what is it. EDI is defined as the exchange of routine business documents in a structured format between computer applications within and between companies. Routine business documents include purchase orders, shipping notices, invoices, and other documents that are now used to conduct business. EDI replaces these paper formats with their electronic equivalents.

The purpose of EDI is not to eliminate paper, but rather to eliminate the time and data entry associated with paper. A widely accepted statistic claims that 70 percent of one computer's business output becomes a second computer's input. In a paper environment, this results in the same information being rekeyed several times into both computers. EDI links the computer processes so that duplicate data entry is not necessary any more.

6.3.1
Reasons for EDI

EDI use in the United States and worldwide has been steadily increasing since its introduction in the late 60s. In 1980, approximately 2.000 U.S. companies were using EDI. By 1990, the number of EDI-capable companies in the U.S. had increased to 22.000. EDI is currently used in over 50 industries including automobile, pharmaceutical, grocery, health care, and manufacturing; and the list is growing. In the period between 1989 and 1992, annual growth of registered EDI users never fell below 70%. Today, over 25.000 users are participating in EDI activity in North America, with another 15.000 participants around the globe. EDI activity is occurring in all major industries at both federal and state government levels, and in companies of all sizes. EDI participants include organisations from small grocery stores to the world's largest shipper, the Department of Defence. A continuation of this steady and significant growth is predicted based upon plans of major companies and their suppliers and customers to incorporate EDI.

Initially, EDI use was limited to large businesses in the grocery, retail and transportation industries. Companies such as General Motors (GM), Boeing and General Electrics (GE) are examples of EDI pioneers. These large organisations became involved in EDI due to their business with hundreds of suppliers. More than smaller companies, these mega-businesses were realising the inefficiencies and redundancies of paper document systems and interchanges. For these companies, the move to EDI capability proved extremely profitable. Another reason that large companies were among the EDI pioneers is that, in order to become EDI-capable, businesses must convince their trading partners to do the same. This often proves to be a difficult task among smaller companies with fewer available resources and funds. Often, large companies, on the other hand, have the clout and financial resources necessary to require EDI of their trading partners.

Many organisations simply give their suppliers the EDI mandate: "Either become EDI-capable or we'll take our business elsewhere". For the most part, smaller companies enter the EDI arena out of fear of losing the business of a big customer or supplier to a competitor who agrees to cooperate with the larger organisation's request to use EDI.

A different example, however, is the approach Mobile Oil took to EDI implementation. Mobile's corporate EDI strategy was to make it as easy and inexpensive as possible for both its customers and suppliers to conduct business electronically with Mobile. To facilitate the achievement of Mobile's vision for EC, they made both PCs and Mobile's customised EDI software 100 percent reimbursable for eligible distributors. Thus, Mobile basically funded the transition to EDI for its distributor base. Their effort to fund EDI for its distributors reveals that Mobile saw EDI as profitable and critical to its ability to maintain market share and gain new business. As EDI use increases in these large businesses and trickles down to others, the ability of organisations to conduct business without EDI will decrease. Already, many companies are finding that they must use EDI in order to maintain their business relationships with other organisations in the industry. EDI participants have been putting pressure on their trading partners to join the movement towards EDI since the time EDI was first introduced.

The pressure from customers to implement EDI is a significant factor driving its growth. In fact, a 1986 study of EDI pioneers revealed that half of the companies surveyed believed that EDI would eventually become a vendor selection criterion. Customer or supplier request accounted for 55 percent of user conversions to EDI; other reasons for this transition include:

- quick access [5%];
- cost savings [7%];
- accuracy of data [8%];
- competitive advantage [12 %];
- improved customer service [12%];
- all these reasons [1%].

In other words, smaller companies believed that sometime in the future, the purchasing policy of their customers would be "no EDI, no business". Additionally, according to a study of 1000 companies the most commonly cited reason for implementing EDI was strong customer demand.

6.3.2
Benefits of EDI

Several benefits of EDI have already been discussed in the comparison of electronic document interchanges to the traditional paper exchanges of business information. The benefits of EDI certainly do not stop here. The following are additional benefits that result from EDI implementation:

- increased business opportunities
- reduced inventory
- more accurate records and decision-making information
- lower data entry costs
- decreased postal mailing costs
- greater customer satisfaction
- reduced order time
- better cash management

6.3.2.1
Increased Business Opportunities

As large organisations move toward EDI capability, they will require the same of their trading partners. Many smaller organisations have found that a move to EDI resulted in a larger market for their goods and services. Mobile Oil (discussed earlier) is an example of one company that saw EDI as a gateway to market stability and new business.

EDI also increases business opportunities and creates a level playing field for organisations who are suppliers of goods and services to the DOD and other government agencies. Thousands of requests for quotation (RFQs) for small purchases that were previously only found on the bulletin boards outside the government contracting office will be made available to small businesses. Posting the RFQs in an electronic form enables businesses to quickly review, select, and respond to the RFQs and eliminates days of administrative processing. The vision for EDI in government procurement is for vendors to know, at the push of a button on their PC, what the selling opportunities are anywhere in the government. EDI has significantly improved the contracting process. Now small companies located in remote areas can compete for business from their PCs without ever leaving their offices. It has provided a wider competitive market.

6.3.2.2
Reduced Inventory

Inventory costs are another expense that can be reduced through the use of EDI. EDI minimises inventory costs by improving a number of factors that contribute to the need for inventory. The level of inventory that must be held to satisfy demand depends primarily upon demand usage, order cycle or lead time, and the uncertainty associated with the two. In a paper-based system, suppliers need to maintain a stock of a particular good so that it may be easily accessed and sent once an order is received. EDI can facilitate a reduction in inventory in two ways:

- By reducing the transaction time, which reduces the lead time, which in turn may reduce safety stock.
- By reducing uncertainty in lead time, providing faster and more accurate information about demand.

Thus, EDI allows businesses to take a proactive, rather than reactive, approach to supplying goods to their customers in a timely manner. For many companies, the reduction in inventory costs is the most significant benefit of EDI. Under a traditional paper-based system, it takes six days for a typical order to arrive at the manufacturer and another four days for the merchandise to be delivered. Using EDI, however, the six days of mailing and processing time are eliminated.

Therefore, only enough inventory to cover the four days of delivery time needs to be carried.

6.3.2.3
More Accurate Records and Decision-Making Information

It has been estimated that even professional typists and data entry clerks miskey two percent of the time. Assuming that 98 percent of the information entered on company business forms such as shipping notices and purchase orders is correct, the remaining 2 percent that is not accurate can be embarrassing and costly. Examples of mistakes made too often are the misspelling of a frequent customer's name, an invoice authorised for a $1,000 payment instead of a $100 payment, or an order to ship 100 items rather than 10 items. Considering the number of business transactions made daily by a typical company, there is great room for error. Add to this the fact that this information can be manually re-keyed as many as 22 different times in the cycle.

EDI ensures greater information accuracy by exchanging data directly between computer systems. This eliminates the need for manual re-entry, thus reducing errors and increasing accuracy. After a company enters data by keying it directly into its system, EDI software edits the data to ensure accuracy. EDI software displays an error message for inconsistencies such as an invalid account number or part number, or an incorrect price. Following the implementation of EDI, a major organisation transmitted 600,000 freight bills electronically in a span of 18 months with absolutely no errors. The elimination of errors alone paid for the cost of developing its EDI system.

6.3.2.4
Lower Data Entry Costs

The most inefficient and cost and labour intensive part of traditional paper-based systems is the manual entering of data from one computer printout into another computer system. Not only is it inefficient because of the high probability for error, it is a burden on company costs and time. As the old adage goes, "Time is money". EDI provides a link for the direct transmission of data from one computer to another. It eliminates the need for duplicate data entry and the costs associated with the performance of these activities. EDI users have reported that they have accurately transmitted invoices within minutes and processed them immediately without any human intervention. Reduced labour processing costs resulting from more efficient data entry allows labour to be used more effectively in problem solving and other activities that require human judgement.

6.3.2.5
Decreased Postal Mailing Costs

Sending a paper order through a postal service is costly and inefficient for several reasons:

- There are time delays associated with mailing and other forms of physical transmission. Mailing by postal services may take anytime from two days to a week.
- There is no acknowledgement of receipt other than an effort made by either the sending or receiving party.
- Once the costs of typing the order and addressing the envelope are added to the cost of postage, a single order can cost five dollars or more. Overnight mailing adds another five to ten dollars to the cost.

Because EDI involves the direct computer-to-computer exchange of information, traditional postal services are unnecessary. The time delays associated with mailing and postal services are eliminated. Value Added Networks (VANs) which function as electronic post offices, provide 24-hour service and can immediately forward the "package" to the recipient's mailbox. Messages can be sent and received with little, if any, lag time. Additionally, EDI software automatically generates and sends back a functional acknowledgement every time an EDI message is received. Thus, in all circumstances, the sender knows with certainty that the message has been received. Finally, EDI eliminates costs associated with paper processes and data entry. With EDI, a business transaction or group of transactions are sent in an electronic envelope through a VAN to the electronic mailbox of a trading partner. Because the forms already exist in an electronic format, there is no need for duplicate data entry. This significantly reduces mailing and handling costs. Many organisations justify EDI implementation by the savings in this area alone.

6.3.2.6
Greater Customer Satisfaction

EDI provides overall greater customer satisfaction by enabling the production of better quality goods, in an efficient manner, in less time. With an efficient EDI system, an order can be received, processed, and shipped almost as quickly as it can be transmitted. The order is sent electronically, received the same day, and the goods are shipped immediately. EDI also provides organisations in industries such as transportation the capability to supply their customers with status reports about their shipments. The ability to locate shipments more quickly adds to customer service and satisfaction. Companies may also use EDI to ease the burden of tedious and often time-consuming responsibilities such as purchasing office supplies, furniture, clothing, and other items not used directly in their production processes.

6.3.2.7
Reduction in Order Time

As alluded to previously, the period of time between sending and receiving an order by mail can often consume a week or more of valuable time. Because EDI eliminates one day of handling time on both ends, in addition to the two to three days in the postal system, the order time is reduced by almost a full week. In many cases, EDI provides the capability for goods to be ordered and shipped the same day.

6.3.2.8
Better Cash Management

By taking advantage of EDI, companies can ensure the purchase of the proper materials at the right time. This enables them to better plan cash disbursements. When EDI is used to transmit an invoice or advanced shipping notice for use in an Evaluated Receipt Settlement (ERS) system, the invoice is handled with consistency, and no guesswork is needed to know when it will be paid. This consistency allows for much more efficient cash management. Additionally, goods are received faster; invoicing and payments are quicker and more accurate, corporate balance sheets are up-to-date; and there is better scheduling of workloads within the receiving department.

Having shown the immense advantages and benefits of EDI, the following chapter explains the technical aspects of EDI.

7 Standards for EDI Documents

As mentioned above it is very important to exchange the information included in an EDI message in a format that can be easily understood by all TPs, regardless of the differences in computer systems. Because communication in EDI is computer-to-computer, rather than person-to-person, the data must be exchanged in a standard format. This means that the information must be contained in some pre-established, uniform arrangement that can be read and deciphered by computers without human intervention. Some organisations have proprietary EDI systems in which they have their own "standard" format used in transferring business information but if they want to exchange data with TPs having a different format they will have problems.

EDI standards are a very important component of EDI. They were first introduced in the 50s and 60s when various industries began to exchange business information electronically. The Transportation Data Coordinating Committee (TDCC) was formed in the late 60s to create standards for transportation industries, such as rail, motor, air and ocean shipping. The success of these standards in the transportation industry led to the creation of standards by TDCC for other industry groups like grocery, chemical and warehousing. However, it was not long before the limitations of these industry-specific standards were realised.

Standards in the United States have since shifted from these company- or industry-specific formats to more general, common formats. Early electronic data interchanges that relied upon proprietary formats for information interchange and were agreed upon by two trading partners could not be applied to other industries or even to another trading relationship. The disadvantages of programming the widely varying formats required by different partners soon became cumbersome and time- and labour-intensive.

In the 1960s, industry groups such as grocery, transportation and chemical began a cooperative effort to develop EDI industry standards for purchasing, transportation and financial applications. However, many of these standards supported only inter-industry trading. Others, such as standards for bills of lading, purchase orders and invoices, were applicable across several industries. As EDI use grew to encompass the business activities of diverse organisations, the idea of national, cross-industry standards began to receive substantial support.

EDI standards, similar to a language, consist of a grammar (syntax, rules for structuring data) and vocabulary of words (data directory, message directory).

These must be agreed between the communicating parties for any message to make sense. Obviously, this agreement could be made individually for any particular communication, but there is a risk of a "Tower of Babel" arising where each communication uses a different language.

Therefore a number of common grammars and vocabularies have been agreed which can be used in a wide variety of situations. The most common examples are

- **ANSI X12:** The formerly most often used EDI standard in the United States;
- **TRADACOMS:** The UK version of the United Nations Guidelines for Trade Data Interchange (UNGTDI);
- **UN/EDIFACT:** The international standard produced by United Nations Joint EDI Working Group.

In recent times Trading Data Communications (TRADACOMS) users are generally migrating to UN/EDIFACT and ANSI X12 will soon be superseded by UN/EDIFACT. It is likely that in a few years UN/EDIFACT will be the single international EDI standard and therefore it is described here in more detail.

7.1
United Nations Electronic Data Interchange for Administration, Commerce and Trade (UN/EDIFACT)

This standard of EDI documents comprises a set of internationally agreed syntax standards, directories and guidelines for structuring and exchange of data that can be generated in character format between independent computer systems. The UN/EDIFACT rules are published in the United Nations Trade Data Interchange Directory (UNTDID).

The most important element of this is the statement of the UN/EDIFACT syntax rules (ISO 9735) for structuring data messages and has been ratified as a European Standard (EN 29735) as well. In ISO 9735 there is a distinction between batch EDI and interactive EDI (I-EDI). Batch EDI is defined as electronic data interchange where no strong requirements exist for formalised telecommunication association between the parties using query and response. Interactive EDI, on the other hand, is defined as the exchange of predefined structured data within a dialogue, which conforms to the syntax defined within the standard for some business purposes, between a pair of cooperation processes, in a timely manner.

ISO 9735 consist of the following parts:

- ISO 9735-1-Syntax rules common to all parts and the syntax service directories
- ISO 9735-2-Syntax rules specific to batch EDI
- ISO 9735-3-Syntax rules specific to interactive EDI

- ISO 9735-4-Syntax and service report message for batch EDI (message type—CONTROL)
- ISO 9735-5-Security rules for batch EDI (authenticity, integrity and non—repudiation of origin)
- ISO 9735-6-Secure authentication and acknowledgement message (message type—AUTACK)
- ISO 9735-7-Security rules for batch EDI (confidentiality)
- ISO 9735-8-Associated data in EDI
- ISO 9735-9-Security key and certificate management message (message type—KEYMAN)
- ISO 9735-10-Security rules for interactive EDI

7.1.1
Batch EDI

7.1.1.1
Batch EDI Interchange Structure

The service string advice (= an optional string used at the beginning of an interchange to specify the service used in the interchange in more detail), if used, and the header and trailer service segments (excluding those used for security and associated data, which are defined in related parts of this International Standard), shall appear in a batch EDI interchange in the order shown below.

Name	Tag	Name
Service String Advice	UNA	Conditional
Interchange Header	UNB	Mandatory
Group Header	UNG	Conditional
Message Header	UNH	Mandatory
Message Body		
Message Trailer	UNT	Mandatory
Group Trailer	UNE	Conditional
Interchange Trailer	UNZ	Mandatory

In the diagram above, the lines to the left show the pairing of header and trailer segments. For simplicity, an interchange containing only one group and one message is shown.

NOTE - Segments for use in UN/EDIFACT messages are defined in the
United Nations Trade Data Interchange Directory (UNTDID).

7.1.1.2
Batch EDI Message Within an Interchange

The following figure shows the possible parts of a Batch EDI interchange.

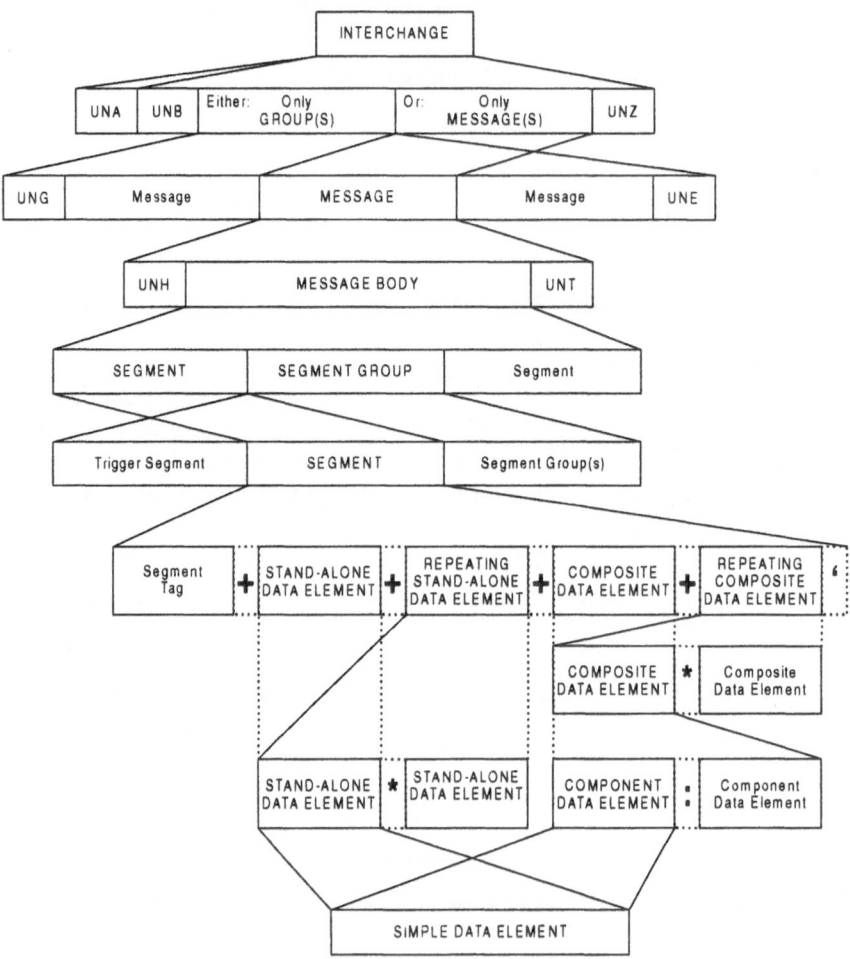

Fig. 7.1. Batch EDI message within an interchange

7.1.2
Interactive EDI

7.1.2.1
I-EDI Interchange Structure

The service string advice (if used) and the header and trailer service segments shall appear in an I-EDI interchange in the order shown below.

Name	Tag	Name
Service String Advice	UNA	Conditional
Interactive Interchange Header	UIB	Mandatory
Interactive Message Header	UIH	Mandatory
Message Body		
Interactive Message Trailer	UIT	Mandatory
Interactive Interchange Trailer	UIZ	Mandatory

7.1.2.2
I-EDI Message Within a Transaction

The following figure shows the possible parts of an Interactive EDI interchange:

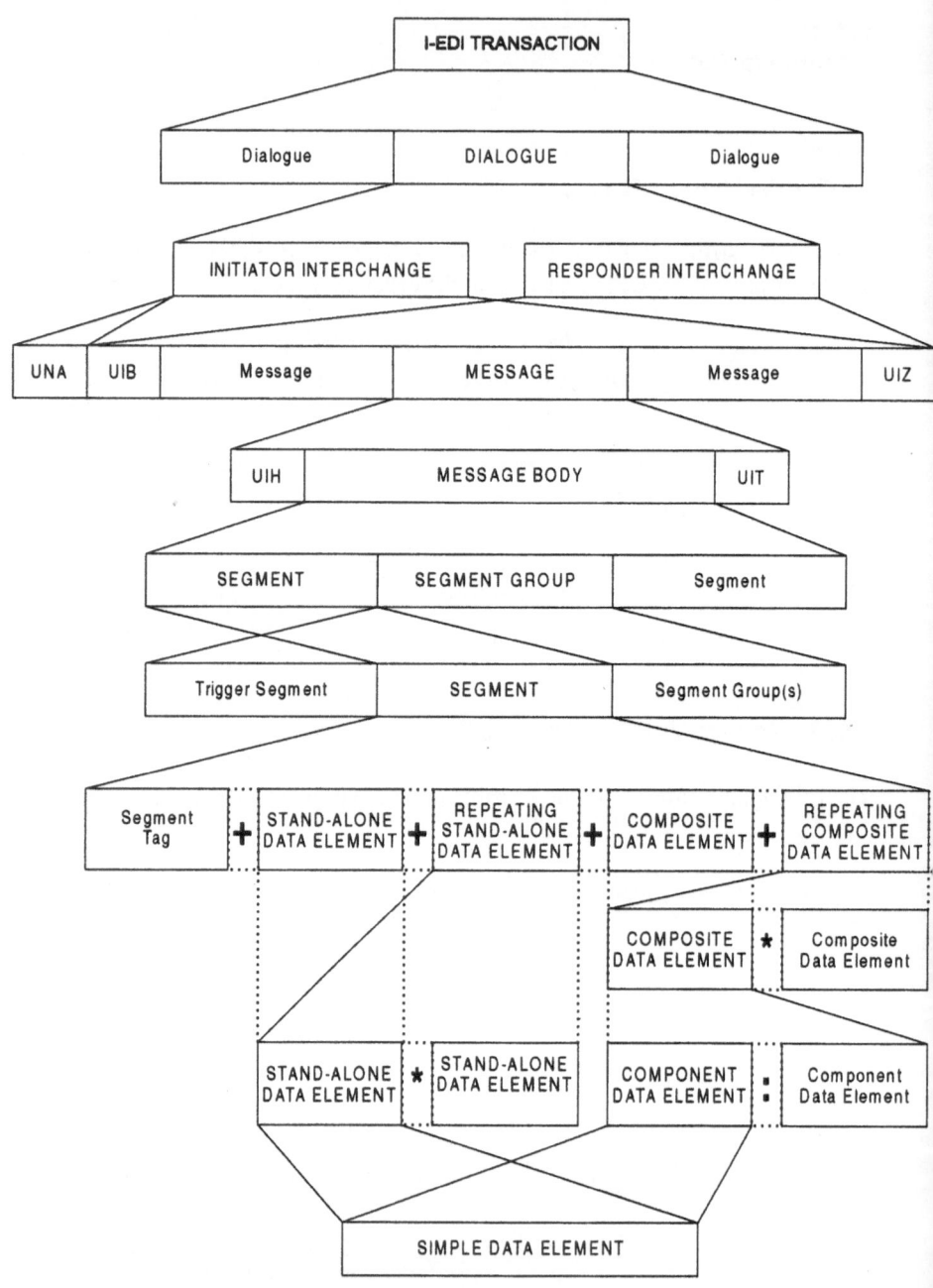

Fig. 7.2. Interactive EDI message within an interchange

7.1.3
Elements of EDIFACT

Although an existing data directory, called Trade Data Element Directory (UNTDED) ISO 7372, is the basis for EDIFACT, a new data directory has nevertheless been developed for each new syntax. Because a harmonised directory is strongly demanded, ISO has established a Data Element Coordination Group (DCG) to correspond to this fact. Furthermore the compelling need for standardised message types that fulfil the requirements of the international trade was realised. Hence the single user need not develop his own message format relying on the basic-syntax rules.

According to the figures above EDIFACT defines different types of blocks:

- Data Elements
- Codes
- Composite Data Elements
- Segments and Messages

Data Elements possess codes or values. Messages are grouped into functional groups (messages of the same type) and then combined to interchanges (set of messages possessing the same originator/recipient identification).

7.1.3.1
Data Elements

EDIFACT defines application data as data elements. Individual data elements can be identified by an unambiguous numerical code within UNTDED (ISO 7372). This standard defines nine different categories of data elements. Numerical and alphanumerical representation is possible and the possible minimal and maximal values are specified as well.

7.1.3.2
Codes

UNTDED specifies explicit code values for each data element. Code sets additionally specify the whole range of possible values for each data element. United Nations/Economic Commission for Europe (UN/ECE) manages a code directory in addition to each data element directory for EDIFACT messages.

7.1.3.3
Composite Data Elements

Some data elements are composed of single data elements, called Component Data Elements that are, if they are grouped, characterised as Composite Data

Elements that have their own reference identification. Furthermore Composite Data Elements are more and more characterised in a more general way and they possess additional qualifying attributes. The definition of special code sets for these attributes enables context-sensitive semantics for those data elements.

7.1.3.4
Segments

In EDIFACT it is possible to group elements together into segments. Segments possess a mnemonic reference within a directory that is part of a segment-identifier and that is leading the segment in the encoded message. By segment definitions each data element within a segment is assigned either a mandatory or a conditional status. If there are already assignments of a conditional status to data elements within a segment, it is possible to change this from a conditional status to a mandatory one.

7.1.4
UN Standard Messages

With the aid of the EDIFACT syntax rules it is possible to build an unlimited number of different messages that is only limited by the way of selecting data elements from the UNTDED. Assuming that it is possible to define a core set of messages that correspond to the different user requirements, especially in the area of international commerce, the UN/ECE has developed the concept of United Nations Standard Messages (UNSMs). The main idea of this concept is to define for each kind of business a very general message that can be used for a whole sector. For example, the form of a bill will vary to a very high degree. The aim is to include all possible functions in the general message for bills, so that only a subset is needed for each special business sector. To reach this goal, the UN/ECE Working Party 4 chose a regional approach for the standardisation of the messages. This means that regional groups are working together and for each special function, only one message is standardised worldwide. This message is certainly a very general one, and as mentioned earlier, only a subset will really be used.

7.1.5
UNTDID—A Collection of EDIFACT Directories

The UNTDID consists of four single directories:

- the Data Elements Directory, a subset of UNTDED ISO 7372;
- the Code Directory for data elements that are defined as codes; they are essential as qualifier in Component Data Elements;

- the Composite Data Element Directory, consisting of combined data elements that themselves consist of Component Data Elements;
- the Segments Directory, consisting of segments that consist of combinations of the data elements mentioned above.

An additional part of the UNTDID consists of a listing of UNSMs.

7.2
Differences and Mutualities Amongst the Different EDI Standards

The differences amongst the several commonly used standards (for example, EDIFACT, ANSI X12, TRADACOMS) consist primarily in differences amongst the separator characters, the definition of segments and the definition of EDI messages.

All the standards include the concept of a special segment that is used to start an interchange. This segment is commonly called header segment. The key data elements contained in this segment are common to most EDI standards, and it is these data elements that are of most interest to Pedi, because these data elements include some information that can be used for addressing, selective retrieval and routing.

An EDI interchange is just a character string. If EDIFACT or TRADACOMS are used as EDI syntax standards, the EDI interchange consists of ASCII characters. If ANSI X12 is used as the EDI syntax standard, the EDI interchange may consist of EBCDIC characters.

8 Transportation of EDI Messages

UN/ECE does not limit the way EDIFACT messages are exchanged between two or more applications. It rather allows each reliable communication service to be used. This implies that there are not only standardised but also proprietary transaction protocols in use. A common property of all of these protocols is that they behave like a file transfer as shown in the next figure:

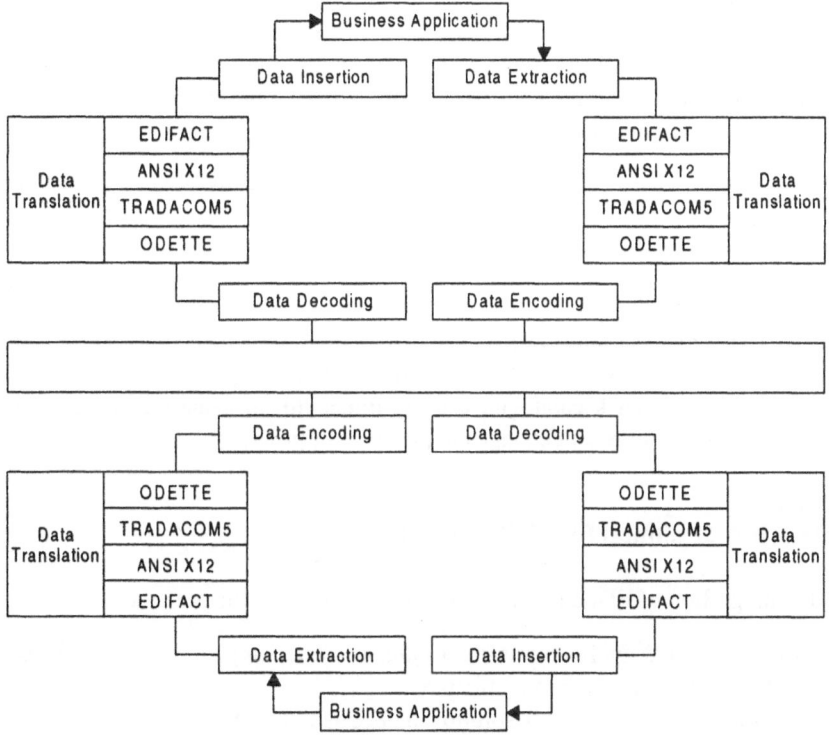

Fig. 8.1. Basic functionality of EDI software

The originating application calls its associated message translator to create a message that is afterwards sent to the specified recipient application by using a specific syntax. Before the message is received at the recipient application it is first translated into a format known by this application.

Generally there are two different transfer possibilities that can be used:

- real-time file transfer using File Transfer and Management (FTAM), another OSI standard;
- store-and-forward mechanism, e.g. X.400 or SMTP.

The most suitable one is certainly X.400.

8.1
EDI Message Transfer via Store-and-Forward Mechanisms

As messaging systems mature, companies are realising the advantages of uniting the worlds of human-readable e-mail and machine-readable EDI messages. Organisations are seeking to leverage their existing infrastructures into broader EC capabilities.

E-mail, of course, is not a replacement for integrated, machine-readable EDI communications. But an integrated EDI/e-mail solution has many potential applications, most of which enable the company to improve customer service and increase competitive advantage:

- An EDI purchase order can be linked with an e-mail message to a sales rep so the rep can follow up on an order;
- an EDI request for information can be linked to human-readable CAD/CAM graphic, avoiding the delays of mail service and the possible illegibility of fax transmission;
- an EDI advance ship notice can be linked to an e-mail message to loading dock personnel, alerting them about the incoming shipment;
- an EDI price/sales catalogue can be linked to an e-mail update, notifying customers about recent changes in products or prices.

8.1.1
Benefits of Linked EDI/E-Mail Messaging

The integration of EDI and e-mail offers multiple benefits to the user:

- In a combined application e-mail enhances and supplements EDI and makes EDI communications more effective.
- An integrated EDI/e-mail solution makes it much easier to track, audit, and respond to EDI communications.
- Although some companies use fax transmissions to supplement their EDI communications, e-mail is easier and cheaper to use. Unlike fax messages, e-mail goes directly to the addressee's computer, is more readily answered, and almost always legible.
- An integrated EDI/e-mail solution allows a company to extend e-mail capability beyond its four walls. This encourages those within the organisation

to take advantage of external messaging, reaching out to and connecting with other business through the Internet or through private e-mail provider.

Because of the needed quality of service for EC and EDI, which can today only be ensured by VANs, the use of the X.400 protocol is certainly the best transfer mechanism for EDI documents. That is why this way of transferring EDI documents is described here in more detail.

8.2
EDI Message Transfer via X.400

There are three different X.400 protocols that can be used to transfer EDI documents:

- the P0 protocol
- the P2 protocol
- the P35 (Pedi) protocol

8.2.1
Using the P0 Protocol to Transfer EDI Documents

This approach is mainly used in the US, and was outlined by the National Institute for Standards and Technology (NIST). In X.400 an integer is generally used to define the content type of an MTS envelope. In this case NIST recommended to use 0 to represent the messages which contain EDI interchanges for an "undefined" content type and therefore the protocol is called P0. Effectively this approach makes use of the P1 protocol to transmit the message. Therefore the approach is often called P1/0 approach. The following figure shows this approach.

A very important drawback of it is that it relies on MHS 1984 and therefore cannot make use of the added extensions of the 1988 standard and the MS. Furthermore the EDI interchange is transparent to X.400; therefore selective retrieval based on EDI attributes is not possible.

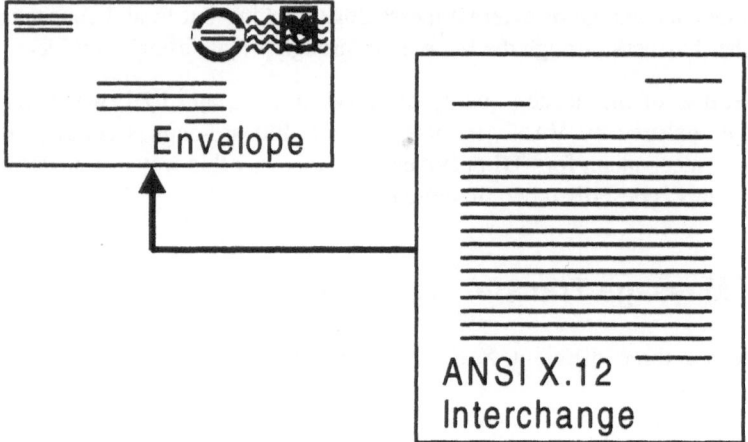

Fig. 8.2. Usage of the P0 protocol for transferring an EDI interchange

8.2.2
Using the P2 Protocol to Transfer EDI Documents

The P2 approach predominates in Europe and is defined by the Trade Data Interchange Systems (TEDIS) guidelines and approved by CCITT and the European Commission. This approach to EDI over X.400 was influenced by the practical approach to EDI by European companies. In many cases these companies have been using proprietary mail systems for EDI; therefore it seemed logical to incorporate EDI at the IPMS level. The TEDIS guidelines simply recommend that the interchange is transported as the text part of an interpersonal message content. As the following figure shows, in this approach the body of the IPM contains the EDI interchange, coded as IA5 string.

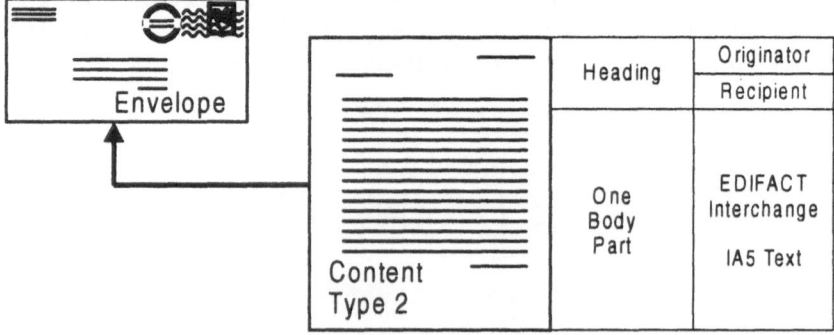

Fig. 8.3. Usage of the P2 protocol for transferring an EDI interchange

8.2.3
Using the P35 (Pedi) Protocol to Transfer EDI Documents

Because of the fact that neither of the approaches mentioned above is ideal and in both approaches many EDI requirements are left unsatisfied, the CCITT developed two new recommendations—F.435 and X.435—which respectively describe the service and the system protocol for an EDI content type. X.435 is alternatively known as Pedi or P35. In contrast to the P0 and P2 approach, Pedi provides a message structure that is specifically designed for EDI interchanges.

8.2.3.1
EDI Messaging System Model

F.435/X.435 describe the EDI messaging system (EDIMS). This is analogous to the IPMS, and is another way of using the generic MTS.

As can be seen in the following figure, the EDIMS includes User Agents (EDI-UA) and message stores (EDI-MS). It optionally includes a physical delivery access unit (PDAU) to permit the delivery of EDI documents to recipients via an existing physical delivery system (PDS). The EDIMS may also include other AUs, e.g. a FAX AU.

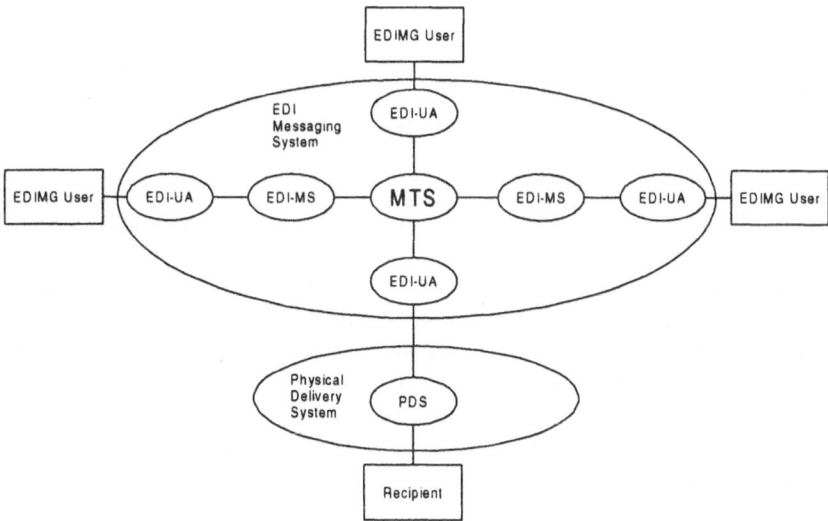

Fig. 8.4. EDI messaging system model

8.2.3.2
EDI Messaging Environment

It is also necessary to consider the environment in which EDI messaging will take place. It is not only necessary for the Pedi content type to control the transfer of EDI responsibility, and to enable the inclusion of EDI document fields in the message header, but it should also be remembered that existing EDI communities generally base their document syntaxes on either EDIFACT, ANSI X12 or UNGTDI. As a matter of fact to ensure interoperability all three must be supported.

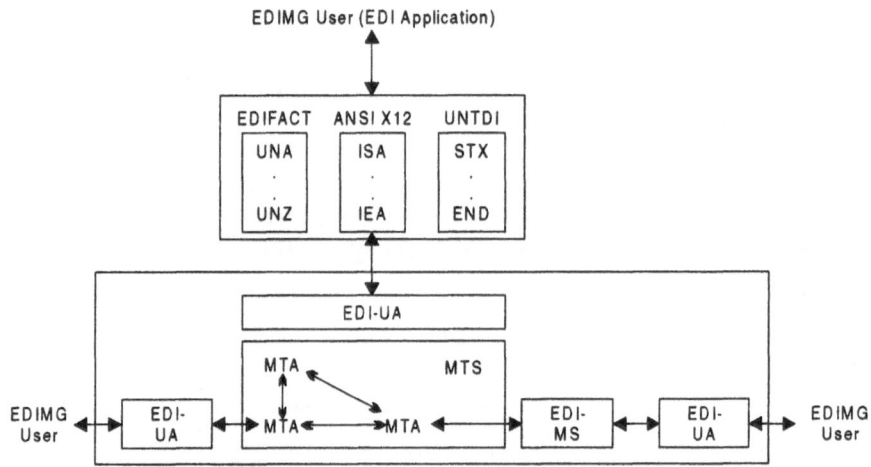

Fig. 8.5. EDI messaging environment

The figure above illustrates the information flow in the EDI messaging environment. EDI users will create their EDI documents and attachments, if any, according to the agreed specification of their EDI community. Using the EDI-UA, the EDI documents and attachments will be submitted as a Pedi message content, and addressed to its recipient(s). The EDI message will then be transferred through the MTS for delivery to a recipient EDI-UA, or a recipient's EDI-MS. The recipient EDI users will then receive a copy of the original EDI document and attachments, via their own EDI-UA, either on delivery, or by fetching the document from the EDI-MS.

8.2.3.3
EDI Message Structure

The structure of a Pedi message is similar to a P2 message and consists of a message header followed by the message body. The message body may consist of one or more body parts, but unlike P2, there is one designated primary body part. This is always an EDI document or a forwarded EDI document, and there may be only one primary body part per message. The remaining body parts contain attachments to the EDI document and may be any recognised external body part type. Note that the message header will contain fields specific to Pedi as well as copies of fields from the EDI document. This is to enable these fields to be made available a EDI-MS attribute.

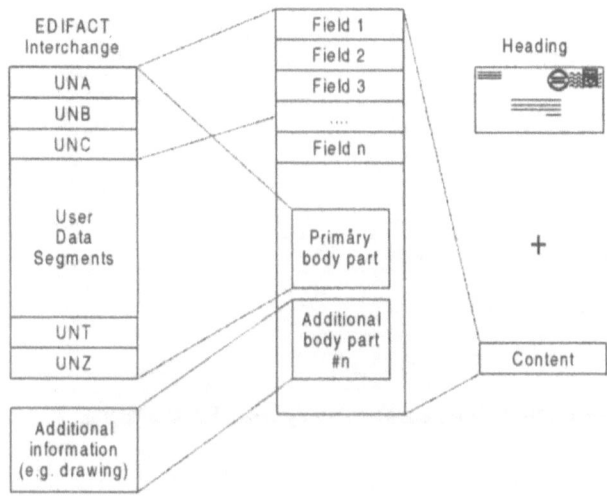

Fig. 8.6. EDI message structure

8.2.3.4
EDI Notifications

As with P2, Pedi also defines notifications. However, while P2 has only two notifications - receipt notification and non-receipt notification—Pedi has three different notifications:

- the negative notification;
- the forwarded notification;
- the positive notification.

All of the three notification types are necessary to support the passing of EDIM responsibility.

The structure of the EDI notifications follows the P2 model, and consists of a set of common fields, followed by fields specific to the notification type.

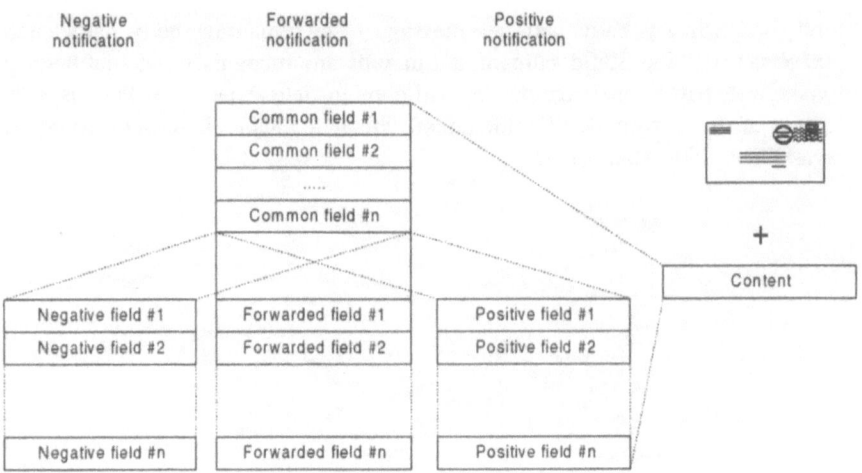

Fig. 8.7. EDI notifications

8.2.3.5
EDI Message (EDIM) Responsibility and Forwarding

The EDIMS includes a concept called EDI responsibility. The need to support the concept of EDIM responsibility is one of the principal reasons why a new content type is necessary for the exchange of EDI documents using MHS and it is the key to the description of EDI notifications and forwarding.

The purpose of introducing the concept of EDIM responsibility is primarily to provide a method for confirming the passing of messages amongst EDI-UAs. EDIM responsibility may also apply to AUs in certain cases.

EDIM responsibility indicates that the EDIM is made available by the EDI messaging (EDIMG) user by the receiving EDI-UA. It is a notional token that may be passed with a message and that can be accepted or refused. EDIM responsibility shall always be accepted when the EDI-UA adds or removes body parts when forwarding. An EDIM cannot leave the EDIMS unless EDIM responsibility has been accepted. If requested by the originating EDI-UA, the recipient EDI-UA, and possibly intermediate EDI-UAs (if requested), shall send EDINs to the originating EDI-UA.

When an EDI-UA receives an EDIM, it shall, if requested, inform the originating EDI-UA that the recipient EDI-UA has accepted or refused EDIM responsibility by sending an appropriate EDIN. If notifications are requested, only one of the following scenarios will happen:

- The EDI-UA accepts EDIM responsibility and therefore sends a positive notification;
- the EDI-UA refuses EDIM responsibility and therefore sends a negative notification;

• the EDI-UA forwards EDIM responsibility and therefore sends a forwarding notification.

In the last case it shall create the appropriate heading fields in the forwarded EDIM.
Body parts that are forwarded cannot be changed anyway. If EDIM responsibility is forwarded, the forwarded EDIM cannot be changed as well. If EDIM responsibility is accepted, body parts may be removed from or added to the original EDIM when creating the forwarded EDIM. Body parts that are removed when forwarding are replaced with place holders to indicate what type of body part was removed. EDIM responsibility forwarding is limited to only one recipient.
It should be noted that EDIMG includes mechanisms to prevent looping when forwarding.

Auto-Actions. ITU-T Recommendation X.413 specifies the concept of an auto-action. This is an action which the MS performs automatically on behalf of the UA. The UA must of course instruct the MS to perform specific auto-actions for a specific set of messages.
The UA can specify a selection criterion based on the MS attributes and instruct the MS to perform a certain auto-action for all messages satisfying the selection criterion.
The auto-actions defined in Recommendation X.413 are not generally useful for Pedi, therefore Recommendation X.435 defines the following auto-actions which are specific to EDI:

• forwarding with responsibility not accepted;
• forwarding with responsibility accepted.

Forwarding with responsibility not accepted auto-action. When the UA instructs the MS to perform this auto-action, it instructs the UA to always forward EDI responsibility for an EDIM that it forwards. That is, the UA instructs the MS to forward all EDIMs that satisfy the certain selection criterion based on the MS attributes, and, in addition, it instructs the MS to generate FNs for the forwarded EDIMs, if any are requested, and to forward the notification request in the EDIM to the recipient of the forwarded EDIM.
The MS always forwards the EDIM unchanged, that is, no body parts are added or dropped, and it forwards the EDIM to at most one recipient.

Forwarding with responsibility accepted auto-action. When the UA instructs the MS to perform this auto-action, it instructs the UA to always accept EDI responsibility for an EDIM that it forwards. That is, the UA instructs the MS to forward all EDIMs that satisfy the certain selection criterion based on the MS attributes, and, in addition, it instructs the MS to generate PN EDINs for the forwarded EDIMs, if any are requested.

If no EDINs are requested, the MS can still auto-forward the EDIM, and to do so it uses the rules for the forwarding with responsibility accepted auto-action.

The MS always forwards the message unchanged, that is, no body parts are added or dropped. The EDIM may be forwarded to more than one recipient.

Registration of auto-actions. If an auto-action is desired, it must be registered, and the selection criteria for the auto-action must be specified.

Since very general selection criteria can be specified for each registered auto-action, it can happen that more than one auto-action is requested for a particular EDIM. However, there are some priority rules specified in Recommendation X.435 that are used in order to decide which auto-action to perform if several are requested. These priority rules can be summarised as follows:

* Look for the "forwarding with responsibility accepted" auto-action and perform as requested. Several auto-actions can be performed by the same MS for the same EDIM, resulting in multiple forwarding.
* If no "forwarding with responsibility accepted" are requested for the EDIM, look for a "forwarding with responsibility not accepted" auto-action and perform as requested. At most one of these auto-actions can be performed by the same MS for the same EDIM, so in this case multiple forwarding cannot take place.

Forwarding and secondary distribution. In EDIMG it may be desirable to receive EDI messages at a central EDI-UA, with subsequent forwarding to the final EDI-UAs. Such a practice would, for example, enable large organisations to perform centralised functions such as logging, auditing, etc. on all EDI message traffic entering that organisation. After the performance of these functions the traffic would be distributed to the EDI UAs serving the recipient EDI applications. Similarly, a VANS provider might operate an intermediary stage as well on behalf of its customers.

Since an intermediate EDI-UA will generally not be the final EDI-UA, there is a need to provide end-to-end confirmation of EDIM responsibility acceptance for an EDIM within EDIMG. The element of service "EDI notification request" allows an originator to request from each recipient positive, negative and forwarded notifications. Together with protocol elements defined in X.435, the "EDI notification request" allows intermediate EDI-UA to indicate, in a forwarded message, whether or not EDIM responsibility has been accepted.

These tools allow EDIM responsibility acceptance to be deferred until an EDIM reaches the final EDI-UA, and provide indication to that EDI-UA that a notification is to be returned to the original originator.

In order to illustrate the use of an EDI-UA as an intermediate stage, three cases are described below. In all three cases, an EDIM originates in EDI-UA 1 and terminates in EDI-UA 3. EDI-UA 2 is the intermediate EDI-UA. In case 1 and 2 it is assumed that the EDIM is forwarded with content unchanged. In all three cases it is assumed that EDI-UA 1 has requested notifications.

Case 1: No forwarding

The figure below shows the scenario.

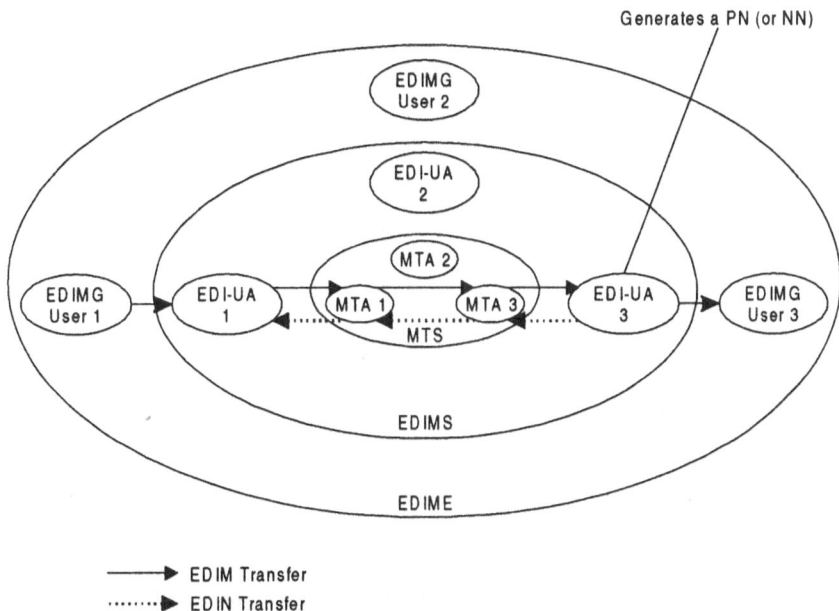

Fig. 8.8. Transfer of EDIM responsibility without forwarding

In this scenario, the EDI document is submitted to the MTS via EDI-UA 1. It is then passed through the MTS and delivered to its recipient EDI-UA 3. If a delivery report has been requested by the originator, it will be generated at EDI-UA 3 and returned to EDI-UA 1. EDI-UA 3 then passes the EDI document and attachments, if any, to the EDI user. EDI responsibility will either be accepted by EDI-UA 3 on behalf of its user, or it is refused. In the first case, a positive notification is sent if requested by the originator of the EDIM; in the latter case, a negative notification is sent back to EDI-UA 1. Note that the meaning of acceptance and refusal of EDI responsibility is dependent on the agreed interpretation specified by the EDI community under which the exchange takes place.

Case 2: Forwarding EDIM responsibility without changing the content

In the second case the EDI message is received by an intermediate EDI-UA and the message together with the EDI responsibility are forwarded to its eventual recipient.

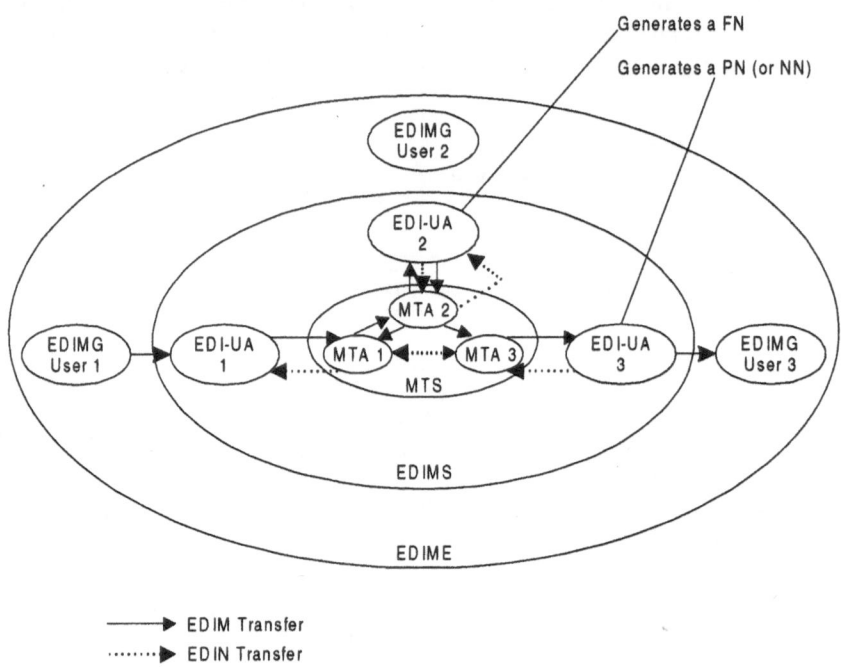

        ───────▶ EDIM Transfer
        ·······▶ EDIN Transfer

Fig. 8.9. Forwarding EDIM responsibility without changing the content

In this scenario, the EDI document is again submitted to the MTS via EDI-UA 1. It is the passed through the MTS and delivered to its recipient EDI-UA 2. If the originator of the EDI message has requested a delivery report, the report is generated and returned to EDI-UA 1.

During the inspection of the message EDI-UA 2 determines that some or all of the EDI messages should be forwarded to EDI-UA 3, along with EDI responsibility. It then creates a new EDI message, encapsulating the original, with or without all of its body parts, and submits the new EDI message to the MTS for transfer to the forward recipient. EDIM responsibility is handled as follows:

When EDI-UA 2 forwards EDIM responsibility, it shall create the forwarded EDIM so that the requested EDINs are received by EDI-UA 1. The following EDINs may be sent:

- If EDI-UA 1 requested notification of forwarding of EDIM responsibility, EDI-UA 2 shall send a forwarded notification (FN) to EDI-UA 1. This EDIN is sent when EDI-UA 2 successfully submits the EDIM to MTA 2.
- If EDI-UA 2 receives a non-delivery notification from MTA 3 (via MTA 2), it may send a negative notification (NN) to EDI-UA 1. Note that EDI-UA 2 has the choice to send or not to send the EDIN in this case.
  No other EDINs may be requested or sent. For example, EDI-UA 2 cannot request notifications from EDI-UA 3, EDI-UA 3 cannot send EDINs to EDI-UA 2.
  In the case of non-delivery, EDI-UA 2 may attempt to resubmit the EDIM to the intended recipient. In this case the NN to EDI-UA 1 is sent only when EDI-UA 2 determines that it shall no longer attempt to resubmit the EDIM to EDI-UA 3.
- If forwarding succeeds, EDI-UA 3 shall send an appropriate EDIN to EDI-UA 1, accepting or refusing EDIM responsibility.

The following should be noted:

- EDI-UA 1 will usually receive several EDI notifications, if it requested a forwarded notification;
- EDI-UA 1 may receive EDI notifications in a sequence other than that in which they were created;
- EDI-UA1 may receive no EDI whatsoever, even if it requested forwarded notification (for example, in the case of catastrophic failure of EDI-UA 2 after MTA 2 has delivered the EDI message to EDI-UA2).

It is EDI-UA 1's responsibility to handle correctly the three points listed above.

Case 3: Forwarding EDIM responsibility with changing the content
This scenario provides for the case where the EDI message prepared by EDI-UA 1 is addressed to EDI-UA 2, and EDI-UA 2 accepts EDIM responsibility for the message prior to forwarding to EDI-UA 3. This would occur, for example , if EDI-UA 2 were to add or remove body parts when forwarding (changes of content). When EDIM responsibility is accepted ED-UA 2 sends a positive notification to the originator, and creates the forwarded EDI message so that no further EDI notifications are received by EDI-UA 1 (the originator). As in the former case, EDI-UA 1 addresses the EDI message to EDI-UA 2, and as in the previous two cases, EDI-UA 3 represents the final destination.

Upon retrieval of the EDI message, EDI-UA 2 returns a positive notification to EDI-UA 1, which means that it has accepted EDIM responsibility. The message is then forwarded to EDI-UA 3. Since initial EDIM responsibility has now been accepted by EDI-UA 2, it has the decision whether to request EDI responsibility, as desired. If requested, the resulting EDIM responsibility relationship shall apply between EDI-UA 3 and EDI-UA 2, i.e. not end-to-end as in the previous cases. In the scenario described here, EDIM responsibility is assumed to be requested, with the result that EDI-UA 3 responds to EDI-UA 2 with an appropriate notification.

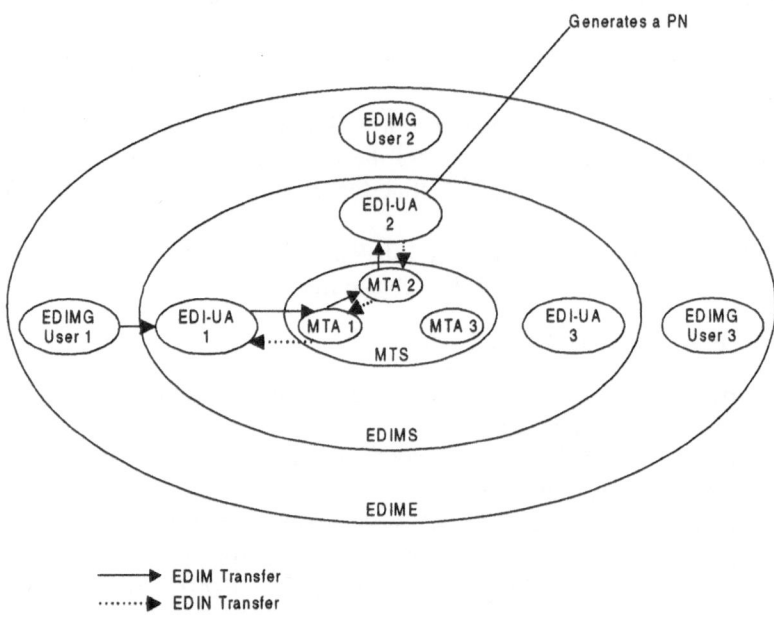

Fig. 8.10. Forwarding EDIM responsibility with changing the content (Part 1)

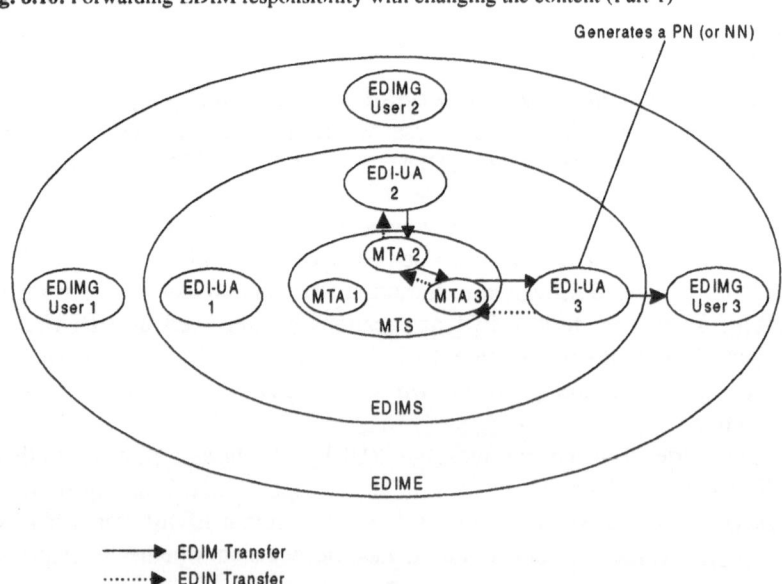

Fig. 8.11. Forwarding EDIM responsibility with changing the content (Part 2)

8.2.3.6
The EDI Messaging System and Physical Delivery

A physical delivery access unit (PDAU) may be used to deliver EDI messages to recipients that do not have EDIMS capability, or for which physical delivery is more appropriate. A PDAU is used when the recipient's address field is specified by a postal O/R address, either by an originator, or as a result of a directory look-up.

While it is up to the PDAU to physically render and subsequently deliver an EDI message sent to it, it is important to note that a PDAU can never accept EDIM responsibility for an EDI message. If an EDI notification request is present, two possibilities exist for the PDAU. If it determines that it can render the EDI message for physical delivery, then it returns a forwarded notification to the originator of the EDI message. In the second case, where it determines that it cannot render or deliver the EDI message, it returns a negative notification to the originator.

When an EDI message is delivered to a physical recipient, then two possibilities also exist for notifying the acceptance of EDI responsibility, and both may be performed for the same message. Firstly, an acknowledgement of receipt may be returned over the physical delivery system to the PDAU; secondly, the PDAU then generates a positive notification on the recipient's behalf, and returns it to the originator solely through the agency of the physical delivery system.

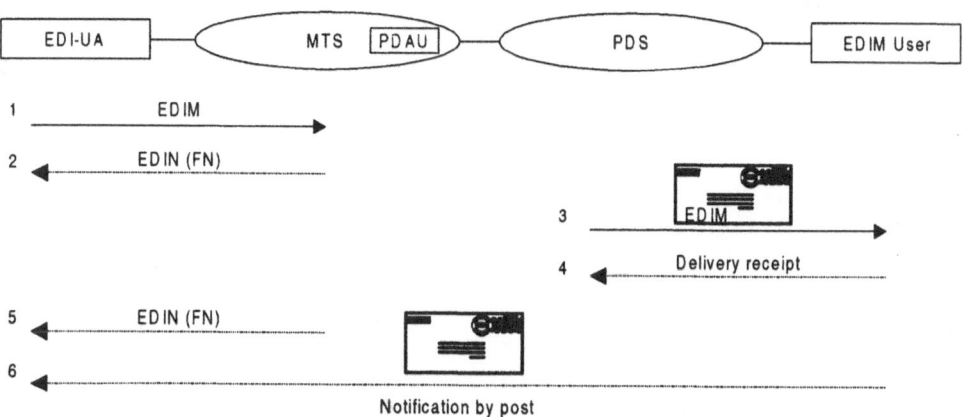

Fig. 8.12. EDIMS and physical delivery

9 Vulnerabilities and Security Requirements of EDI Messaging Environments

9.1 Vulnerabilities

Because in an EDIME data are transferred between two or more users (user applications) via a network, there are several possible vulnerabilities. Some of these vulnerabilities are not unique to sending EDI messages but to sending data in networks. The following vulnerabilities are the most important ones the security service must provide solutions for:

- Masquerade
- Message sequencing
- Message loss
- Modification of information
- Repudiation
- Leakage of information
- Manipulation of information by EDIMG users

9.2 Masquerade

Masquerade occurs when an entity successfully pretends to be a different entity and can take place in a number of ways. An unauthorised MTS-user may impersonate another to gain unauthorised access to MTS facilities or to act to the detriment of the valid user, e.g. to discard his messages. An MTS-user may impersonate another user and so falsely acknowledge receipt of a message by the "valid" recipient. A message may be put into the MTS by a user falsely claiming the identity of another user. An MTS-user, MS, or MTA may masquerade as another MTS-user, MS, or MTA.

Masquerade threats include the following:

- impersonation and misuse of the MTS;
- falsely acknowledge receipt;
- falsely claim to originate a message;
- impersonation of an MTA to an MTS-user;

- impersonation of an MTA to another MTA.

A masquerade usually consists of other forms of attack and in a secure system may involve authentication sequences from valid users, e.g. in replay or modification of messages.

9.3
Message Sequencing

Message sequencing threats occur when part or all of a message is repeated, time-shifted, or re-ordered. This can be used to exploit the authentication information in a valid message and resequence or time-shift valid messages. Although it is impossible to prevent replay with the MHS security services, it can be detected and the effects of the threat eliminated.

Message sequencing threats include the following:

- replay of messages;
- re-ordering of messages;
- pre-play of messages;
- delay of messages.

Users should not assume that EDIMs shall be delivered in the correct sequence. EDI applications should be able to recover from duplication and out-of-sequence messages, provided that MHS offers protection against the modification of information while messages are within the MHS environment.

9.4
Message Loss

Vulnerability to message loss is considered critical in the EDIMG environment.
Two types of message loss are distinguished:

- catastrophic failure of an EDI-UA, EDI-MS or MTA,
- loss of individual message(s).

EDIME users and service providers may need to consider more carefully issues concerning transfer of messages between EDIM responsibility domains:

- from the originating EDI-UA user domain;
- between relaying domains;
- to the recipient EDI-UA user domain.

9.5
Modification of Information

Information for an intended recipient, routing information, and other management data, may be lost or modified without detection. This could occur to any aspect of the message, e.g. its labelling, content, attributes, recipient, or originator. Corruption of routing or other management information, stored in MTAs or used by them, may cause the MTS to lose messages or otherwise operate incorrectly.

Modification of information threats include the following:

- modification of messages;
- destruction of messages;
- corruption of routing and other management information.

9.6
Repudiation

Repudiation can occur when an MTS-user or the MTS may later deny submitting, receiving, or originating a message. Repudiation threats include the following:

- denial of origin;
- denial of submission;
- denial of delivery.

It should be noted that repudiation vulnerability in the EDIME is considered to be critical. Such vulnerability may be increased by the use of certain MHS services (e.g. auto-forwarding, redirection).

9.7
Leakage of Information

Information may be acquired by an unauthorised party by monitoring transmissions, by unauthorised access to information stored in any MHS entity, or by masquerade. In some cases, the presence of an MTS-user on the system may be sensitive and its anonymity may have to be preserved. An MTS-user other than the intended recipient may obtain a message. This might result from impersonation and misuse of the MTS or through causing an MTA to operate incorrectly. Further details on the information flowing in an MTS may be obtained from observing the traffic.

Leakage of information threats include the following:

- loss of confidentiality;
- loss of anonymity;
- misappropriation of messages;

• traffic analysis.

9.8
Manipulation of Information by EDIMG User

The EDI community has additionally identified a further vulnerability where the integrity of a message content is altered subsequent to EDI interchange (i.e. by either or both of the originating EDI-UA and recipient EDI-UA). This vulnerability includes manipulation of message content in the originator's local store after non-repudiation of submission and/or manipulation of message content in the recipient's store after non-repudiation of delivery.

9.9
Security Requirements

Because of the very important and personal data being transferred in an EDIME, several security requirements have to be fulfilled in order to guarantee the needed level of privacy and integrity of the transferred data and to counter the vulnerabilities listed above. The security requirements can be summarised in a few words:

• authentication;
• data confidentiality;
• data integrity;
• non-repudiation.

9.9.1
Authentication

If authentication is guaranteed, the recipient(s) of a message can be sure that the sender is the user it pretends to be.

9.9.2
Data Confidentiality

If data confidentiality is guaranteed, the recipient(s) of a message can be sure that the content of the message has not been disclosed to users other than the intended recipient(s).

9.9.3
Data Integrity

If data integrity is guaranteed, the recipient(s) of a message can be sure that the content of the message has not been tampered with during the transmission.

9.9.4
Non-repudiation

If non-repudiation is guaranteed, the recipient(s) of a message can be sure that neither the transmitted data had been manipulated, nor that the received message has not been sent by the specified sender.

Since an X.500 directory is very often used in conjunction with a messaging environment, the required security services can be obtained form the strong authentication mechanism specified in Recommendation X.509.

10 Cryptography and Key Management

10.1
Cryptography

When a packet is routed from the source to the destination, it is routed through many intermediate nodes. It is at the intermediate nodes where anyone at one of the machines that handle the packets could reassemble the packets that make up the EDI Interchange and could therefore read it, copy it, alter it, or delete it. In the case where an EDI Interchange is carried using the SMTP protocol, the case could arise where the message cannot be delivered to the final recipient, so that the message must be stored at an intermediate node. Once again, the message is susceptible to any number of security threats mentioned above.

Encryption plays an essential role in protecting the privacy of electronic information against possible security threats from a variety of attackers. The goal of encryption is to turn otherwise readable text into something that cannot be read, and therefore understood. By making the text unintelligible, the encryption discourages anyone from reading or copying the EDI Interchange while it is in transit between the trading partners. Encryption conveys confidentiality to the EDI Interchange.

Encryption is based on two components: an algorithm and a key. An algorithm is a mathematical transformation that takes plain text or other intelligible information and changes it into unintelligible cipher text. The inverse mathematical transformation, back to the original from the cipher text is also done, and called decryption. In order to encrypt the plain text, a key is used as input in conjunction with an encryption algorithm. An algorithm can use one of any of a large number of possible keys. The number of possible keys each algorithm can support depends on the number of bits in the key. If the key length is n, then there are 2n possible key combinations, with each different key causing the algorithm to produce slightly different cipher output. For example , if the key length is 40, then there are 240 results in 1,000,000,000,000 possible key combinations.

An encryption algorithm is considered "secure" if its security is dependent only on the length of its key. Security cannot be dependent on the secrecy of the algorithm, the inaccessibility of the cipher or plain text, or anything else—except the key length. If this were true about a particular algorithm, then the most efficient—and only—attack on that algorithm is a brute-force attack, whereby all key combinations must be tried in order to find the one correct key.

10.1.1
Symmetric Encryption—Secret Key Cryptography

Encryption algorithms whereby two trading partners must use the identical key to encrypt and decrypt the EDI Interchange are called symmetric encryption algorithms. That means that, if an EDI Interchange is encrypted with one key, it cannot be decrypted with a different key. The key used in most symmetric encryption algorithms is just a random bit string, n bits long. These keys are often generated from random data derived from the source computer.

The use of symmetric encryption simplifies the encryption process, each trading partner does not need to develop and exchange secret encryption algorithms with one another (which incidentally would be a near-impossible task). Instead, each trading partner can use the same encryption algorithm, and only exchange the shared, secret key.

There are however drawbacks with "pure" symmetric encryption schemes; a shared secret key must be agreed upon by both parties. If a trading partner has n trading partners, then n secret keys must be maintained, one for each trading partner. Symmetric encryption schemes also have the problem that origin or destination authenticity (non-repudiation of origin, and receipt) cannot be proved. Since both parties share the secret encryption key, any EDI Interchange encrypted with a symmetric key, could have been sent by either of the trading partners.

The following algorithms are the most commonly used symmetric encryption algorithms:

- Data Encryption Standard (DES);
- Triple DES;
- RC2 and RC5;
- International Data Encryption Algorithm (IDEA).

10.1.1.1
Data Encryption Standard (DES)

The most widely used symmetric encryption algorithm is DES. It is widely used in the banking industry for Electronic Funds Transfer (EFT). Furthermore it is a U.S. government encryption standard. It is also in the public domain, which means anyone can implement the algorithm, including those in the international community. DES was primarily designed for, and is used for bulk encryption of data. DES is prohibited by the U.S. government for export.

DES has been analysed by cryptographers since the middle of the 1970s and its security is considered as "known": in other words, the security of DES is dependent on the length of its keys, and estimates can be provided for the time and the effort required to derive the DES key from a known 8 byte plain text/cipher text pair. DES specifies a 56 bit key, so 256 keys are possible.

10.1.1.2
Triple-DES

Triple-DES is a variant on DES that encrypts the EDI Interchange 3 times, with 2 independent 56 bit keys, giving an effective key length of 112 bits. This makes a brute-force attack on triple DES, which means trying every single key, not feasible. DES and Triple-DES actually can be implemented in 3 different modes. It is recommended that both algorithms should be used in Cipher Block Chaining (CBC) mode, which gives added protection by making the cipher text blocks dependent on each other, so that changes in the cipher text can be detected as well.

10.1.1.3
RC2 and RC5

RC2 and RC5 are proprietary symmetric algorithms of RSA Data Security, Inc. Both algorithms are, unlike DES, variable key length algorithms. As a matter of fact by specifying greater or lesser key length, these algorithms can be configured to provide greater or lesser security. U.S. government restrictions limit RC2 implementations to 40 bits when exported outside the United States. RC2 should be used in CBC mode as well.

10.1.1.4
International Data Encryption Algorithm (IDEA)

This algorithm is an iterated block cipher with a 64-bit block size and a 128-bit key size and was published in 1991. The key length of IDEA is over twice that of DES and is longer than triple-DES. The IDEA algorithm is patented in both the United States and abroad. The IDEA algorithm in CBC mode is used by Pretty Good Privacy (PGP), a popular electronic mail encryption protocol, for encryption. Individual users of PGP have a royalty free license to use the IDEA algorithm.

IDEA is a newer algorithm and has not been studied as much as DES. It has a more than adequate key length, and PGP supports a configurable key length from 384 to 2048 bits. Indications are that IDEA is a secure algorithm and its use in PGP makes it the most widely used encryption algorithm for Internet electronic mail.

By using what is called public key cryptography, the management of symmetric keys can be simplified to the point whereby a symmetric key can be used not only for each trading partner, but also for each exchange between trading partners. In addition, public key cryptography can be used to unambiguously establish non-repudiation of origin and receipt.

10.1.2
Asymmetric Encryption—Public Key Cryptography

Public key cryptography is based on the concept of a key pair consisting of a public key and a private key. Each half of the pair (one key) can encrypt information that only the other half (one key) can decrypt. The key pair is designated and associated to one, and only one, trading partner. One part of the key pair (the private key) is only known by the designated trading partner; the other part of the key pair (the public key) is published widely but is still associated with the designated trading partner.

The keys are used in different ways for confidentiality and digital signature. Both confidentiality and digital signature depend on each entity having a key pair that is identified only with them and each party keeping one pair of their key pair secret from all others.

Digital signature works as follows: Trading Partner A uses her secret key to encrypt only a part of a message or the whole message, then sends the encrypted message to Trading Partner B. B gets Trading Partner A's public key (anyone may get it) and attempts to decrypt the encrypted part of Trading Partner A's message. If it decrypts, then Trading Partner B knows it is from A—because only A's public key can decrypt a message encrypted with A's private key, and A is the only one who knows her private key.

Confidentiality works as follows: Trading Partner A would retrieve Trading Partner B's public key, and encrypt the message with it. When Trading Partner B receives the message, she would decrypt the message with her private key. Only her private key can decrypt information that was encrypted with her public key. In other words, anything encrypted with B's public key can only be decrypted with B's private key. Since public key encryption algorithms are considerably slower than their symmetric key cousins, they are generally not used for the encryption of whole EDI Interchanges. It is estimated that software encryption using DES is 100 times faster than software encryption using RSA (a public-key encryption algorithm). Hardware encryption using DES is anywhere from 1,000 to 10,000 times faster than hardware encryption using the RSA asymmetric encryption algorithm. Instead of being used for bulk encryption, public-key encryption algorithms are used to encrypt symmetric encryption keys. They are also used as an efficient means of exchanging and managing symmetric encryption keys.

In most real world applications, asymmetric encryption algorithms are therefore not used to encrypt the message or part of the message itself. Furthermore a combination of both techniques is used in order to get the security advantage of public key systems and the speed advantage of secret key systems.

The public key system is used to encrypt the secret key used for encrypting the bulk of a file or a message. Such a protocol is called digital envelope.

10.1.2.1
RSA Public Key Algorithm

RSA is a public key encryption algorithm that has become a de facto standard in its use for symmetric key management. The mathematics of public key cryptography is complicated, but are based on mathematical manipulations of large numerical quantities. In the case of RSA, deriving the private key from the public key is based on the difficulty in factoring large numbers. An RSA public key is generated by multiplying two large prime numbers together, deriving the private key from the public key involves factoring the product of the two large prime number.

Unlike the symmetric encryption algorithms discussed above, the RSA asymmetric encryption algorithm's security is based on the size of the number that needs to be factored. The size or "modulus" of the product of two prime numbers can be factored using some "fast factoring algorithms" which currently exist. The computing power required by these "fast factoring algorithms" can be estimated, and thus the time and cost to factor a number of any given size can also be estimated.

When using the RSA encryption algorithm to encrypt symmetric keys, support of 512 bit to 1024 bit variable key lengths is required. In general, asymmetric algorithms require longer keys to provide the same level of security as their symmetric key cousins. A 512 bit RSA encryption key is equivalent to a 64 bit symmetric key. A 768 bit RSA encryption key is equivalent to an 80 bit symmetric key.

The RSA algorithm is protected by a patent within the United States, but it can be freely used outside.

10.1.3
Conclusion

There are many encryption algorithms that are secure and can provide confidentiality for an EDI Interchange. For most commercial applications a key length of at least 75 bits is recommended. For EDI Interchanges of minimal value, 40-bit RC2 or 56-bit DES are probably adequate. For more valuable EDI Interchanges, the use of Triple-DES, IDEA, or 128 bit length RC2 or RC5, is recommended. In the case of asymmetric encryption for EDI transactions requiring the use of RSA encryption, a 768 bit RSA encryption key should be used. For very "high" value EDI transactions, at least a 1024 bit or higher key is required.

In order to provide confidentiality for EDI Interchanges on the Internet, a standard encryption algorithm (s) and key length (s) must be specified. For interoperability to occur between two trading partners the encryption algorithms and key lengths must be agreed either beforehand, or within an individual transaction.

When choosing public key encryption algorithms, the following criteria should be considered:

- how secure the algorithm is;
- how fast implementations of the algorithm are;
- whether the algorithm is available for international as well as domestic use;
- the availability of APIs and tool kits in order to implement the algorithms;
- the frequency of the use of the algorithm in existing implementations.

Sufficient key lengths must be chosen with regard to the value of the EDI Interchange so that brute-force attacks are not worth the time or effort compared to the value of the Interchange.

10.2
Key Management

10.2.1
Symmetric Keys

The use of symmetric encryption is based on a shared secret. Two trading partners using a symmetric encryption algorithm must be able to do the following:

- generate a random symmetric key and agree upon its use;
- securely exchange the symmetric key with one another;
- set up a process to invalidate a symmetric key that has been compromised or needs changing.

Each trading partner would then need to do this with each and everyone of their trading partners. The management and distribution of symmetric keys can become an insecure and cumbersome process.

Pure symmetric key management schemes also have the problem that origin authenticity cannot be proved. Since two parties share a secret encryption key, any EDI Interchange encrypted with a symmetric key, could have been sent by either of the trading partners—both of whom have knowledge of the key.

As previously mentioned, by using public key cryptography, the management of symmetric keys can be simplified such that a symmetric key can be used not only for each trading partner, but also for each exchange between trading partners. In addition, public key cryptography can be used to unambiguously establish non-repudiation of origin and receipt.

The use of public key cryptography for encrypting the digital envelope simplifies the management of the symmetric keys and makes their exchange much more secure. Trading partners do not need to agree on secret symmetric keys as part of the trading partner agreement, nor is there the origin authenticity problem that is inherent with pure symmetric key management schemes.

A symmetric key can be randomly generated by the software for each EDI transaction between trading partners. Symmetric keys generated on a per transaction basis are sometimes referred to as "session keys". Since a unique symmetric key is generated for each EDI transaction, key maintenance is no longer required. Trading partners do not need to invalidate compromised or expired keys. Each symmetric or "session" key is used only one time.

Additional security is also realised using the method described above; in the unlikely event that one of the symmetric keys is compromised, only one EDI transaction is affected, and not every transaction in the trading partner relationship. Public key encryption also provides a secure way of distributing symmetric keys between trading partners. Since only the receiving trading partner has knowledge of her private asymmetric key, she is the only one that can decrypt the symmetric key encrypted with her public asymmetric key—and is thus the only one who can use the symmetric key to decrypt the EDI Interchange.

To impart confidentiality to an EDI Interchange using public key cryptography for symmetric key management, the following steps would be performed when Trading Partner ABC sends to Trading Partner XYZ:

- The EDI translator outputs the EDI Interchange.
- A random symmetric key of the specified length is generated.
- The EDI Interchange is encrypted using the randomly generated symmetric key with the chosen encryption algorithm.
- The random symmetric key is then encrypted using XYZ's, the receiving trading partner's, public asymmetric key.
- The encrypted symmetric key and encrypted EDI Interchange are then enveloped and sent to the trading partner.

On the receiving side, the following steps would be performed:

- The symmetric key is decrypted using XYZ's private asymmetric key.
- The decrypted symmetric key is then used to decrypt the EDI Interchange.
- The decrypted EDI Interchange is then routed to the EDI translator.

10.2.2
Public and Private Keys

The use of public key cryptography to simplify the management of symmetric encryption keys presents the user with two problems; protecting the private key, and binding a trading partner's identity to his public key. Most likely, the user will never know what his private key is. The software will generate a random private key, encrypt it, and store it in a file or database. The private key is accessed indirectly by the user through access to the software. User access to the software is generally controlled by a password, pass-phrase, and/or certain access rights. These are internal security policies, and are company-specific. It is important to control the access to the private key since any unauthorised access can lead eventually to the revocation of the corresponding public key.

10.2.2.1
Trust and Public Keys

When using public key cryptography, there is a "trust" issue that arises: How can one trading partner be sure that the public key of another trading partner is bound to that trading partner, and is valid? Trading partners must exchange public keys or be able to access each other's public key in a manner that is acceptable to each of the trading partners. One method by which trading partners can exchange public key information is through the use of public key certificates. Public key certificates come in many different formats, and the trust model on which they are based also come with different underlying assumptions.

Public key certificates based on the X.509 standards are however becoming prevalent in their use. The X.509 certificate is the binding of an entity's distinguished name (a formal way of identifying someone or something in the X.500 world, in our case, it would be a trading partner) to a public key. A certificate contains the following:

- **Signature-Algorithm-ID:** An object identifier that denotes the algorithm that was used to sign the certificate.
- **Issuer:** The distinguished name of the certification authority that issued the certificate. The certification authority is typically the organisation responsible for issuing and managing the certificates. The name of the certification authority may correspond to the distinguished name of a DSA associated with that certification authority.
- **Validity:** A predefined period of time during which the certificate is considered to be valid.
- **Subject:** The directory name of the subject of the certificate (that is, the user to whom the certificates pertains).
- **Algorithms:** One or more algorithm identifiers with the public key of the subject of the certificate.
- **Signature:** An asymmetrically encrypted, hashed version of all the above parameters computed by the certification authority that issued the certificate. It is computed by applying the algorithm used to sign the certificate plus the certification authority's secret key.

Certificates have the following properties:

- Any user who has the public key of the certification authority can recover the public key of users served by the certification authority. This is because the public key of the certification authority is used to verify the digital signature which guarantees the integrity of the certificate information held for each user.
- No party other than the certification authority can modify a user's certificate without it being detected.

Certificate issuers are called certification authorities (CA), which can be any trusted central administration willing to vouch for the identities of those to whom it issues certificates and their association with a given key. A company may issue certificates to its employees, a university to its students, a town to its citizens. In order to prevent forged certificates, the CA's public key must be trustworthy: a CA must either publicise its public key or provide a certificate from a higher-level CA attesting to the validity of its public key. The latter solution gives rise to hierarchies of CAs.

Certificate issuance proceeds as follows:

The user requesting a certificate generates its own key pair and sends the public key to an appropriate CA with some proof of its identification. The CA checks the identification and takes any other steps necessary to assure itself that the request really did come from that user, and then sends him a certificate attesting to the binding between that user and its public key, along with a hierarchy of certificates verifying the CAs public key. The user can present this certificate chain whenever desired in order to demonstrate the legitimacy of her public key.

Since the CA must check for proper identification, organisations find it convenient to act as a CA for its own members and employees. There are also CAs that issue certificates to unaffiliated individuals. Different CAs may issue certificates with varying levels of identification requirements. One CA may insist on seeing a driver's license, another may want the certificate request form to be notarised, yet another may want fingerprints of anyone requesting a certificate. Each CA should publish its own identification requirements and standards, so that verifiers can attach the appropriate level of confidence in the certified name-key bindings. CAs with lower levels of identification requirements produce certificates with lower "assurance". CAs can thus be considered to be of high, medium, and low assurance. One type of CA is the persona CA. This type of CA creates certificates that bind only e-mail addresses and their corresponding public keys. It is designed for users who wish to remain anonymous, yet want to be able to participate in secure electronic services.

A certification path is a list of certificates that allows a user to get the public key of another user. Each item in the list of certificates is a certificate for the certification authority of the next item in the list. In this way, a certification path constitutes an unbroken chain of trust among certification authorities.

In the most general case, in order to reciprocally authenticate each other, users must obtain the complete forward-and-return certification paths from either a directory or by local means. That is, they must possess all the public keys of the certification authorities in each other's domain in order to be able to access the public key information signed by those certification authorities. A number of useful optimisations, however, are made to this general scheme. In particular, it can be expected that in most cases, two entities wishing to authenticate each other will share the same certification authority and therefore will not need to rely on certification paths, as both have the public key for the same certification authority.

Where two entities do not share the same certification authority, it is usually simpler for them to exchange, through some off-line mechanism, each other's certification authority public key, as well as the public key of any intermediate certification authorities that may be in the communication path between the two entities. Alternatively, certification paths may be stored in the directory and the two entities will rely on a series of directory queries to obtain full certification paths. When this method is selected, implementations are usually designed to cache the results of directory look-ups for later use.

10.2.3
Conclusion

Since there already exists a trust relationship between EDI trading partners, until the use of certification authorities becomes more established and better profiling is done with X.509v3 certificates, it is recommended that the trading partners "self-certify" each other, if an agreed upon certification authority is not used. In the near term, "self-certification" means that the exchange of public keys and certification of these keys must be handled as part of the process of establishing a trading partnership. Furthermore, great effort should be made such that widespread use of certification authorities relying on X.509v3 certificates is no longer a vision, but a reality.

11 Security Mechanisms for EDI over X.400

Recommendation X.402, §10 provides an abstract security model for message transfer. The security model provides a framework for describing security services that counter potential vulnerabilities within the MTS and between MTS-User to MTS-User. EDIMG vulnerabilities may also be countered by security services which are outside the existing model in Recommendation X.402. The following text describes how the EDIM vulnerabilities are countered using Recommendation X.402 security services, enhanced security services defined in Recommendation X.435, and pervasive mechanisms defined in Recommendation F.435.

11.1
Masquerade

The existing MHS security services which counter this vulnerability are

- message origin authentication;
- secure access management;
- security labelling;
- proof of delivery;
- proof of submission.

Since an EDI-UA/MS is deemed in the MHS architecture as belonging to one user, it is not considered appropriate to provide selective access control for the various operations that may be performed on a EDI-MS. However, there is a requirement for security audit trail to record the actions of the EDIMG user.

11.2
Message Sequencing

The existing MHS security service which counters this vulnerability is message sequence integrity.

This security service has limited effect as it is based on the provision of an integer by the originating EDI-UA with no assurance as to uniqueness or continuation.

It is considered that the MHS environment should not be required to ensure message sequence integrity, but should support detection of sequence integrity failure (by additional provision of audit/logging facilities and/or the provision of third-party notary services). It is recommended to consider it the responsibility of the EDIMG user to recover from sequence errors and message duplication.

11.3
Message Loss

Message loss could occur potentially over any peer-to-peer communications link (e.g. by deliberate malicious act), or by the failure or incorrect behaviour (whether by malicious intent or otherwise) of any MHS component (EDI-UA, EDI-MS, MTA). The following categories of message loss are distinguished:

- catastrophic message loss (i.e. failure of a component);
- loss of individual messages in the EDI-MS—whether malicious or accidental;
- MTS message loss.

11.3.1
Catastrophic Failure

Failure of the EDI-MS is potentially catastrophic and desirably needs some protection, at least in terms of detection. This should be provided by an off-line archive to hold all submitted and delivered messages. It should be mentioned that detection and recovery from message loss using such archive mechanisms is considered to be a local matter.

Failure of any component in the MTS may similarly be catastrophic and can again be protected by an off-line archive of messages. As for the message store, detection and recovery from message loss using such archive mechanisms in the MTS is a local matter as well.

11.3.2
EDI-MS Specific Message Loss

Loss of individual messages in the message store—whether malicious or accidental—shall require the provision of a secure audit trail to enable the detection of such loss. Such a service may need to be provided to the EDIMG user and to EDI-MS management.

11.3.3
MTS Specific Message Loss

Loss of individual messages in the MTS (whether malicious or accidental) shall also require the provision of a secure audit trail to enable the detection of such loss. Such a mechanism would need to be provided on a per-MTA and a per-MD basis depending on the security policy in force. A secure MTA/MTS audit trail could be realised as a pervasive mechanism and is a local issue.

11.3.4
End-To-End Message Loss

The following description assumes that the functionality of the EDI-UA (including any associated components to meet such functionality—e.g. encryption devices) is trusted.

The existing "Message Sequence Integrity" service does not guarantee the detection of message loss, since it relies on the provision of an integer value by the originating EDI-UA. In practice, the effective operation of this service may be achieved with a common code of practice between EDIMG users.

As a result, MHS services which may provide an indication of message loss are confined to services offered to the originating EDIMG user. Whereas the existing "Proof of Submission and Delivery" services provide some degree of confidence that the message has not been lost, they do not operate end-to-end. In particular they do not take account of the scenario where the recipient EDI-UA and EDI-MS are not co-located. There is therefore a requirement for a "Proof of Receipt" (i.e. by the recipient EDI-UA) service. This capability is realised by the user requesting an EDI notification which may be secured. The EDI notification indicating the status of EDIM responsibility as accepted, forwarded or refused includes elements which associates the notification with the subject message.

In an EDIMG environment the proof of receipt may therefore be provided by signing the EDI notification service using the existing MTS security elements. In particular the EDI-UA to EDI-UA security service of "message origin authentication" may be used to sign the EDI notification on submission of the EDI notification to the MTS.

NOTE: This service is called "proof of EDI notification" and/or "non-repudiation of EDI notification" in EDIMG depending on the strength of the mechanism provided.

The MTS mechanism used on message submission to provide this service is defined as the MTS submission abstract operation in Recommendation X.411, § 8.2.1.1.1.28 "Content-integrity-check". In this instance the message content is the EDI notification. Proof of association between the subject message and replying EDI notification is provided by subject message EDI identifier and, if included in the subject message, the message content-integrity-check argument.

11.4
Modification of Information

The existing MHS security services which counter this vulnerability are

• connection integrity;

- content integrity.

These security services provide sufficient protection against the modification of message content. It is also noted that the use of double enveloping (i.e. with encrypted checksum on the outer envelope) may provide additional protection.

NOTE: EDI-UAs are trusted entities in terms of content integrity.

11.5
Repudiation

Services which offer protection against repudiation in the EDIMG environment are fundamentally concerned with formalising the forwarding of EDIM responsibility.

The security services as defined in Recommendation X.402 are

- non-repudiation of origin;
- non-repudiation of submission ;
- non-repudiation of delivery.

These security services only cover some areas of transfer between EDIM responsibility domains, which may be of significance in an EDIMG environment (as illustrated in the following figure). Areas which are not covered by security services provided in 1988 for message handling include:

- between EDIMG user domains (i.e. end-to-end);
- between MTS management domains;
- between an EDI message store and a recipient EDI-UA.

Therefore services and/or pervasive mechanisms defined in X.402 cover the following deficiencies:

- non-repudiation/proof of transfer;
- non-repudiation/proof of retrieval;
- non-repudiation/proof of EDI notification;
- non-repudiation/proof of content.

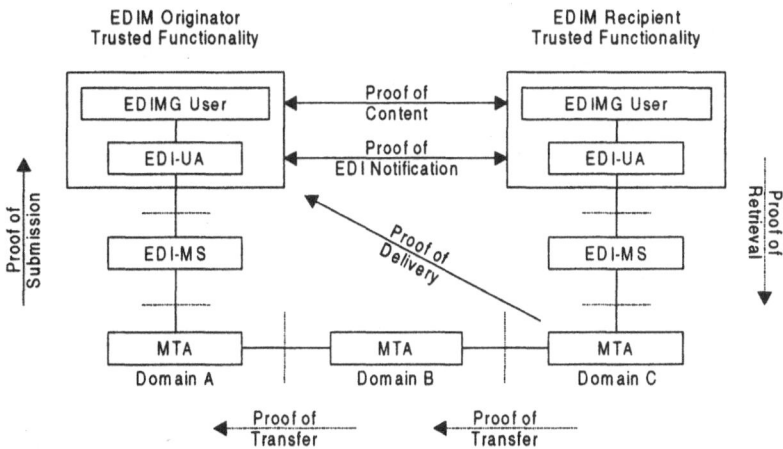

Fig. 11.1. EDI responsibility transfer

"Non-repudiation/proof of transfer" counters the vulnerability of repudiation of responsibility between MTA and/or management domains. EDIMG environments may provide such a service using additional pervasive mechanisms, such as security logs and archives within MTA and/or MTS boundaries. Such pervasive mechanisms provide a "secure MT audit trail" to record the message details and trace information.

"Non-repudiation/proof of retrieval" counters the vulnerability of repudiation of responsibility of a message between a UA and an MS. EDIMG environments may provide such a service using additional pervasive mechanisms, such as security logs and archives within EDI-MSs. Such pervasive mechanisms provide a "secure EDI-MS audit trail" to record EDIMG user actions in the EDI message store.

"Non-repudiation/proof of EDI notification" counters the vulnerability of repudiation of an EDI notification EDI-UA to EDI-UA. This vulnerability may be especially relevant in the case of EDI forwarding, redirection, etc, in addition to the scenario of delivery to an untrusted EDI message store.

Two mechanisms have been defined for non-repudiation of EDI notifications; the first uses the trusted EDI notification as described above, the second uses an external notary system.

"Non-repudiation/proof of content" counters the vulnerability of manipulation of information by the EDIMG user after the message has been received by the EDI-UA. Although such vulnerability is outside the MHS environment, the MHS environment may provide assistance in terms of trusted return of content and notarisation services. There are several ways this requirement may be supported using the secure messaging environment based on the security services provided in 1988.

Firstly non-repudiation of content by the originating EDI-UA may be provided by the existing "Non-repudiation of Origin" security service.

Secondly non-repudiation of content by the recipient EDI-UA may be provided by returning the subject content within the EDI notification and submitting the EDI notification to the MTS using the "Non-repudiation of Origin" security services.

Thirdly, by notarisation services, such services may be achieved by forwarding messages via a mutually trusted third-party notary (i.e. using existing MHS security services).

NOTE: Non-repudiation services (which may imply the involvement of a third party) are considered stronger than "proof-of" services.

11.6
Leakage of Information

The existing MHS security services which counter this vulnerability are

- connection confidentiality;
- content confidentiality;
- secure access management;
- message flow confidentiality.

These security services provide sufficient protection against the leakage of message content. It is also noted that the use of double enveloping could provide some protection against traffic analysis.

NOTE: UAs are trusted entities in terms of content and message flow confidentiality.

11.7
Manipulation of Information by EDIMG User

Manipulation of information by the EDIMG user may be countered by the use of the "Non-repudiation of Content" security service.

11.8
Additional Pervasive Mechanisms

EDIMG may provide additional pervasive mechanisms as follows:

- secure EDI-MS audit trail,
- secure MT audit trail;
- EDI-MS archive;
- MT archive;
- security of MTA management and routing information.

11.8.1
Secure EDI-MS Audit Trail

This facility would monitor and record EDI-UA actions on the message store. It would also provide support for "proof of retrieval".

It is strongly recommended that "secure EDI-MS audit trail" should be controlled via a secure link or other secure local means to protect against masquerade.

11.8.2
Secure MT Audit Trail

This facility would monitor and record all MTA actions. It would also provide additional support for "proof of submission", "proof of transfer", "proof of delivery", security of the administration of the MTA.

11.8.3
EDI-MS Archive

This mechanism is potentially useful for providing recovery from MS failure i.e. by providing a secure off-line archive of all submitted and delivered messages. Detection and recovery from message loss using such archive mechanism is a local matter.

11.8.4
MT Archive

This mechanism is potentially useful for providing recovery from MTA failure i.e. by providing a secure off-line archive of all messages. Detection and recovery from message loss using such archive mechanism is a local matter.

12 Security Mechanisms for EDI over the Internet

The first attempt to define methods for using the Internet for EDI were made in RFC 1767, which defines the packaging of ANSI X12 and UN/EDIFACT Interchanges in a MIME envelope. However, several additional requirements for obtaining multivendor, interoperable services, and how the EDI transactions should be packaged and transmitted, have come to light since the effort concluded. These currently focus on security issues like EDI transaction integrity, confidentiality and non-repudiation in various forms. Standards in these and other areas are necessary to insure interoperability between EDI implementations over the Internet. Various technologies already exist for these additional features, and the primary requirement is to review and select a common set of components for use by the EDI community when it sends EDI over the Internet.

Additional requirements that mimic many of the heading fields found in ITU-T Recommendation X.435 EDI messages (e.g. Interchange Sender, Interchange Recipient, Interchange Control Reference, Communications Agreement ID, and Syntax Identifier) are also needed to support efficient exchanges between the Internet, and Value Added Networks.

The reason why a lot of effort is made to use the Internet for EDI is that traditional VAN connectivity is slow and expensive. On the other side the Internet promises lower costs and is more easily accessible than traditional methods of communication. The predominant problem with the use of the Internet for EDI is interoperability between vendor products, especially in the areas of integrity, confidentiality, digital signature, and non-repudiation. In the Internet Engineering Task Force (IETF) there is a working group called Electronic Data Interchange-Internet Integration (ediint) that has the objective to make the vision of using the Internet for EDI a reality by providing information on how interoperability can be achieved.

12.1 E-Mail Encryption Protocols

Great effort has been made recently to develop mechanism and methods for making mailing over the Internet more secure and to provide a similar quality of service than VANs do. E-mail encryption protocols are intended to protect messages as they are transmitted over the Internet. Various encryption protocols do exist, each of them capable of adequate security. The problem is that they do not only suffer from the lack of good implementations in the sense of transparent e-mail encryption, but they use different encryption algorithms and authentication techniques. Further, they take different approaches to handling such issues as key management—issues that are just as important as encryption in ensuring the safety of data. As a consequence the needed level of multivendor interoperability is not guaranteed. The most important of these protocols are

* Secure Multipurpose Internet Mail Extensions (S/MIME);
* Pretty Good Privacy (PGP);
* Pretty Good Privacy with MIME Extensions (PGP/MIME);
* Privacy Enhanced Mail (PEM);
* MIME Object Security Services (MOSS);
* Message Security Protocol (MSP).

Although e-mail encryption protocols differ in particulars , all of them must support the security requirements mentioned below:

* confidentiality;
* authentication;
* non-repudiation;
* integrity.

To provide these services, all five protocols rely on secret key encryption to encrypt the message, public key encryption to encrypt the secret key, and a hash function to provide authentication.

Confidentiality means that only the indented recipient(s) of a message can read it. To ensure this security service a sender encrypts the message with the public key of the recipient(s), and so only the recipient(s) is (are) able to encrypt the message with its (their) private keys.

Authentication means that it is guaranteed that the message is really from its stated originator, and it can be achieved by applying public encryption in conjunction with a one-way hash function. One-way hash functions have the property that they are easy to compute in the one direction but cannot be reversed. A hash code (also called a message digest) is a small chunk of data unique to a particular message but much shorter than the message itself. Usually the hash codes are 112 to 160 bits long. A hash code is like a digital fingerprint that makes it possible to distinguish one message from another, even if the two differ by only a single bit. To provide authentication, an encryption protocol extracts a hash function from the message, encrypts it with the sender's private key, and appends the encrypted hash code to the message. The hash code can be decrypted by anyone who knows the public key, but the fact that the decryption is successful proves that whoever originated the message had the private key and therefore was the person he or she claimed to be. If a hash code is encrypted with a private key, then not only authentication and **non-repudiation** of origin could be guaranteed, but **message integrity** as well. To provide message integrity, which means that the message has not been tampered with during the transport, the hash code is encrypted with the private key of the sender and then transferred to the recipient. The recipient first decrypts the secret key of the sender together with the hash code. Then he computes the hash code of the message using the same hash function as the sender did. If both hash codes are the same, then the message has not been changed during the transport.

The two most important differences between the e-mail encryption protocols are

- the size of the keys used to encrypt the messages;
- key management.

As far as key management is concerned, for a public key system, there are two aspects:

- finding the public key of any of the communication partners;
- guaranteeing that the public key really corresponds to that party.

Currently there exist two different main approaches for key management:

- certification hierarchies;
- the web of trust.

As the name implies, certification hierarchies take a hierarchical approach to key management; in fact, this model can be thought of as a client-server take on the entire issue.

Certification hierarchies are based on a central certification authority (CA) that is implicitly trusted by all e-mail users. Physically the certification authority can be many things, including dedicated hardware and software running on a server. For instance, the U.S. Postal Service (Washington, D.C.) plans to issue certificates on floppy disks to users who requested them. In practice, however, for most secure e-mail applications, the CA is likely to be a software running on a mail server.

A variation on this idea is the so-called certification chain. In this scheme, a root certification authority signs certificates designating other authorities to certify keys on its behalf. Those authorities in turn can certify other authorities, and so on. At the bottom of the certification hierarchy is a certificate that signs the user's individual public key.

Certification hierarchies tend to work well in structured organisations, such as corporate networks. But for this scheme to be used in communications among different organisations, they must all agree on a certification authority that can be trusted by all, and this is not easy. For example, corporate users are not likely to trust certificates signed by the certification authority of a competing company or by an authority run by governments of hostile countries.

The certification hierarchy most commonly used by secure e-mail schemes is the one defined in X.509. The X.509 specification, which defines the relationship of the certification authorities, is part of the X.500 Series of Recommendations for a global directory structure using distinguished names to track users and other objects stored in a directory.

In the web of trust the users decide for themselves which keys are valid. In other words, a user decides to trust keys that come from a trusted source and arrive via a secure or reasonably secure route that ensures they have not been tampered with. For instance, a worker might trust a key, sent by a known co-worker, that arrives over a secure private network—but not a key purporting to be from an employee from another company arriving via the Internet. Users can also sign the keys of other users, which leads to the web part of the web of trust.

12.1.1
Secure Multipurpose Internet Mail Extensions (S/MIME)

S/MIME provides a standard way to send and receive secure electronic mail and is based on the Internet MIME standard (RFC 2045-RFC 2049). The S/MIME specification consists currently of the following Internet Drafts which have been sent to the Internet Engineering Steering Group (IESG) for consideration as a Proposed Standard. It is an attempt to graft MIME support onto underlying Privacy Enhancement for Internet Electronic Mail (PEM) standards (RFC 1421-1424). It provides mechanisms for encrypting not only plain text, but also for multimedia data. Furthermore it guarantees interoperability between two implementations, which is another important advantage.

One potential drawback of S/MIME is that it permits the use of keys that are too small to ensure adequate security. All S/MIME implementations must include 40-bit RC2. But symmetric encryption algorithms are considered to be adequate for commercial security only with a key length of at least 75 to 90 bits.

However, S/MIME also recommends 56-bit DES in CBC mode and (either 128 or 168 bit) DES EBE3-CBC. The last recommendation especially solves the security problem mentioned above. But work has to standardise the symmetric encryption algorithm and the key length so that adequate security is offered and interoperability between two different implementations is possible.

Another cryptographic weakness of S/MIME is that eavesdropper can distinguish between encrypted and signed and encrypted messages. This violates the principle of disclosing a minimum amount of information.

The most controversial aspect of S/MIME is probably its signature format. A S/MIME signed message is a MIME multipart in which the first part consists of the data to be signed and the second part is a complete Public Key Cryptographic Standard(PKCS) #7 signed message. This protects quite well against munging by mail transport, but has two problems. First, the size of the message is doubled. Second, the fact that two signed messages are identical is not enforced (if it were, mail munging would cause too many signatures to fail).

S/MIME's key management scheme, for example, calls for the hierarchical structure defined in the latest update of X.509 Version 3, the most flexible to date. This explicitly allows certification paths to start in the local security domain (the certification authorities controlled by an organisation) of the public key user system. The other notable feature of X.509 Version 3 is its ability to bind keys directly to Internet e-mail addresses rather than only to X.500 distinguished names.

S/MIME's key management model is in some ways a hybrid between certification hierarchy and the web of trust: As with the web of trust, net managers have to configure each client with the list of trusted keys. In this case, they are the keys of the certification authorities.

Full certificates are included in every S/MIME signed message. This makes messages a few hundred bytes larger and also makes key management a little easier. Whenever a user receives a signed message, the key of the signatory goes into the local key database, where the keys are stored. If the key is signed by a trusted certification authority, trust can be established for that key, and it can be used in the future without remote database operations. Thus, end-users do not have to authenticate the key each time they employ it.

12.1.2
Pretty Good Privacy (PGP)

Pretty Good Privacy has certainly become a de facto standard for e-mail encryption on the Internet. One reason for that might be that PGP's underlying cryptography is quite sound-RSA (up to 2048 bits), IDEA with a 128 bit key and MD5. Another advantage of PGP is that it is packed in a single application (i. e. a single binary) that performs encryption, decryption, signing, verification, and key management, so that it does not depend on a great deal of infrastructure. Furthermore, interoperability amongst implementations is guaranteed.

But nevertheless PGP has some significant drawbacks which makes it still not suitable for fully transparent e-mail encryption. Certainly its main missing feature is the lack of MIME integration. Thus PGP is not suitable for multimedia types other than US-ASCII text.

Contrary to S/MIME, PGP uses the web of trust for managing the keys, which is another big drawback, especially in business settings. Overall, PGP's key management is hard to learn, time-consuming, and dependent on a great deal of manual intervention. Every time a new key is needed, either for checking a signature or encrypting e-mail, users must perform several manual operations. First, they have to get the key; it could be on their partner's web page. Then they must put the key into the local PGP key database, called a keyring in PGP terminology. Next they have to check to make sure the key is valid. In PGP's current implementation, these steps require fairly complex and tricky command lines. But this complexity is not inherent in PGP; key management may eventually become easier.

12.1.3
MIME Security with Pretty Good Privacy (PGP/MIME)

PGP/MIME attempts to add the ability of handling MIME objects to PGP and is specified in RFC 2015. The other advantages and disadvantages of PGP are the same.

12.1.4
Privacy Enhanced Mail (PEM)

Privacy Enhanced Mail (PEM) is a standardised encryption protocol specified in RFC 14221-RFC 1424. And it provides the following security services for text-based e-mail messages:

- authentication;
- integrity;
- confidentiality;
- and non-repudiation.

12.1.4.1
Originator Authentication

In RFC 1422 an authentication scheme for PEM is defined. It uses a hierarchical authentication framework compatible with ITU-T Recommendation X.509. Central to the PEM authentication framework are certificates, which contain items such as the digital signature algorithm used to sign the certificate, the subject's distinguished name, the certificate issuer's distinguished name, a validity period, indicating the starting and ending dates the certificate should be considered valid, the subject's public key along with the accompanying algorithm. This hierarchical authentication framework has four entities. The first entity is a central authority called the Internet Policy Registration Authority (IPRA) acting as the root of the hierarchy and forming the foundation of all certificate validation in the hierarchy. It is responsible for certifying and reviewing the policies of the entities in the next lower level. These entities are called Policy Certification Authorities (PCAs) which are responsible for certifying the next lower level of authorities. The next lower level consists of Certification Authorities (CAs) responsible for certifying both subordinate CAs and also individual users. Individual users are on the lowest level of the hierarchy.

This hierarchical approach to certification allows one to be reasonably sure that certificates coming from users, assuming one trusts the policies of the intervening CAs and PCAs and the policy of the IPRA itself, actually came from the person whose name is associated with it. This hierarchy also makes it more difficult to spoof a certificate because it is likely that few people will trust or use certificates that have untraceable certification trails, and in order to generate a false certificate, one would need to subvert at least a CA, and possibly the certifying PCA and the IPRA itself.

12.1.4.2
Message Confidentiality

Message confidentiality in PEM is implemented by using standardised cryptographic algorithms. RFC 1423 defines both symmetric and asymmetric encryption algorithms to be used in PEM key management and message encryption. Currently, the only standardised algorithm for message encryption is the Data Encryption Standard (DES) in Cipher Block Chaining (CBC) mode. Currently, DES in both Electronic Code Book (ECB) mode and Encrypt-Decrypt-Encrypt (EDE) mode, using a pair of 64-bit keys, are standardised for symmetric key management. For asymmetric key management, the RSA algorithm is used.

12.1.4.3
Data Integrity

In order to provide data integrity, PEM implements a concept known as a message digest. The message digests that PEM uses are known as RSA-MD2 and RSA-MD5 for both symmetric and asymmetric key management modes. Essentially both algorithms take arbitrary length "messages" which could be any message or file, and produce a 16-octet value. This value is then encrypted with whichever key management technique is currently in use. When the message is received, the recipient can also run the message digest on the message, and if it has not been modified in-transit, the recipient can be reasonably assured that the message hasn't been tampered with maliciously. The reason message digests are used is because they are relatively fast to compute, and finding two different meaningful messages that produce the same value is nearly impossible.

However, PEM has the same drawback as PGP according to its limitation to the encryption of only text-based messages. To add mechanisms for encrypting MIME content types like images, audio and video date to the security services of PEM, the Internet specified another encryption protocol called MIME Object Security Services (MOSS).

12.1.5
MIME Object Secure Services (MOSS)

MIME Object Secure Services (MOSS) is another attempt to encrypt multimedia data and is specified in RCF 1847 and RFC 1848. Today, there are no full-fledged commercial implementations available, but there is broad support for adopting some parts of MOSS—especially the multipart/signed message format—for use with other cryptographic protocols. The multipart/signed format allows different parts of a message to be individually encrypted and signed.

MOSS is technically sophisticated and its cryptography is mostly sound. However, the choice of symmetric encryption algorithms (and key length) is left unspecified, so that interoperability between two implementations cannot be guaranteed.

MOSS supports two modes for key management:

- X.509 conformant key management;
- completely manual key management, using the web of trust.

Users get more options, but one feature that is missing is a cryptographic hash of the public key. Without it, users either have to trust the mechanism that delivered the key or examine the entire key to be sure it has not been tampered with. This creates a serious problem.

12.1.6
Message Secure Protocol (MSP)

The Message Secure Protocol (MSP) attempts to refocus the secure e-mail protocols developed by the NSA (National Security Agency) for use with its Defence Secure Data Networking System so that they can be used with Internet MIME mail.

MSP contains two features that are lacking in any of the other proposals, and both could be quite important in a business or, especially, a government context. The first is a cryptographically strong signed-receipt capability (sometimes called a non-repudiation of receipt) which offers proof that the recipient actually received a message and that the message received was identical to the message sent. The second feature is the ability to classify messages as top secret, for example. A MSP mail client rejects messages if the users do not have the appropriate authorisation to read them.

One problem with MSP is that interoperability between two implementations cannot be guaranteed; they may choose different and non-overlapping sets of algorithms. Before communication is possible, users must agree on both the use of the MSP protocol and a specific algorithm set. The two main algorithm suites implemented for MSP are the Mosaic suite (the encryption used with the Clipper chip) and RSA with DES.

A few implementations of MSP are already offered, and several more are in progress. For interactions with the government, this is probably a good choice.

MSP, which closely resembles S/MIME, is based on the X.509 certification hierarchy and raises similar issues about signed message formats. At present it is still unclear whether MSP will be implementing X.509 extensions that make the handling of certification more flexible.

13 EDI Naming, Addressing, and Use of a Directory

13.1 Introduction

When two trading partners are doing EDI today, the EDI relations are commonly the subject of negotiations and agreements between these trading partners. Therefore the use of directory is not critically required for obtaining necessary information on the trading partner, since partners can exchange required information during bilateral negotiations. Nevertheless the use of X.500 directory services has a lot of advantages, especially where security issues are concerned.

The use of directory services will become a necessity as "open" EDI relations become common. An EDI relation said to be "open", if a company opens its EDI gateway to all potential partners, without negotiating individual bilateral agreements with each partner.

Open trading relations are of course the norm in paper-based business transactions. Any company is willing to accept a purchase order from any other company, provided payment is suitably guaranteed.

In the near future it can be expected that companies are willing to accept EDI purchase orders from any originator, provided that payment is suitably guaranteed.

Once open EDI trading relations become common, there will be a significant need for directory services in order to avoid the manual operations required at present to discover the potential partner's O/R address, its EDI capabilities, etc. Additionally, the security services defined in Recommendation X.509 will become a necessity as well.

This clause describes the functions that an EDIMG user may obtain from the directory, if the directory is available to the EDIMG user. If the directory is not available, the functions described here may be performed as a local matter.

13.2 EDI Naming

EDI users (trading partners) identify each other by a "name" which is essentially an arbitrary alphanumeric string. An EDI name is defined to be such an alphanumeric string. EDI standards authorities (for example EDIFACT and ANSI X12) define specific instances of EDI names. The EDI name is normally unique within a particular EDI community, but may not be globally.

The EDI communities may be organised

- by an industry group (e.g. CEFIC, EDIFICE);
- as a private trading group of a large corporation;
- around a third-party EDI service provider.

The EDI names used by any of the above community types are one of the following forms:

- A formalised name issued by an internationally recognised naming authority (e.g. DUNS, EAN) which is globally unique.
- A formalised name issued by a multinational company; the name is unique within the company's trading community and the multinational company acts as a naming authority within this community.
- A free form name assigned by the trading partners themselves, subject only to a uniqueness check by the organiser or operator of the community, acting as a naming authority.

NOTE: All these name forms exist today, and it will take some time for EDIMG users to migrate to globally unique naming.

EDI standards allow the use of a qualifier together with the EDI name. The qualifier identifies the naming authority that assigned or endorsed the alphanumeric string. Globally unique EDI naming is achieved by using EDI names with the appropriate qualifier code.

There is no geographic element in an EDI name, such as country of operation.

The EDIMG users send EDI interchanges with as little addressing information as possible. The EDI name is a static entity which exists for a long period, unlike an address which may change from time to time.

13.3
Suggested DIT Structure for EDI

Annex B of Recommendation X.521 suggests some common naming practices and DIT structures in which locality, country and root can be immediately superior to entries of the object class organisation.

When organisation is immediately subordinate to root it denotes an international organisation. The community types identified above operate internationally, and therefore, the majority of these community types may be classified as international organisations.

A directory structure is suggested in which each community of EDI names is grouped under the organisation that serves as the naming authority for that community (company, industry group, service provider). In this case the entry associated with each EDI name is an alias entry; the actual entry for the EDIMG user is elsewhere in the DIT, as described above.

The following figure illustrates a DIT structure which accommodates the requirements of the EDI community. A new generic object class, EDI user, is created. The attributes in its entry identify the name of the EDIMG user, and to the extent that they are present, the capabilities of the EDIMG user and the attributes of a message handling user as defined in Recommendation X.402.

NOTE 1: The figure illustrates the directory information tree (DIT) that is implied by the object class EDI user that defines the characteristics of an EDI user. The attributes in its definition identify the EDI user's name and, to the extent that they are present, identify the EDI user's capabilities. If the DIT structure cannot be applied to a special organisation, the recommended DIT structure can be changed by defining new object classes including the attributes needed. Additionally it must be specified where these object classes are added to the existing DIT.

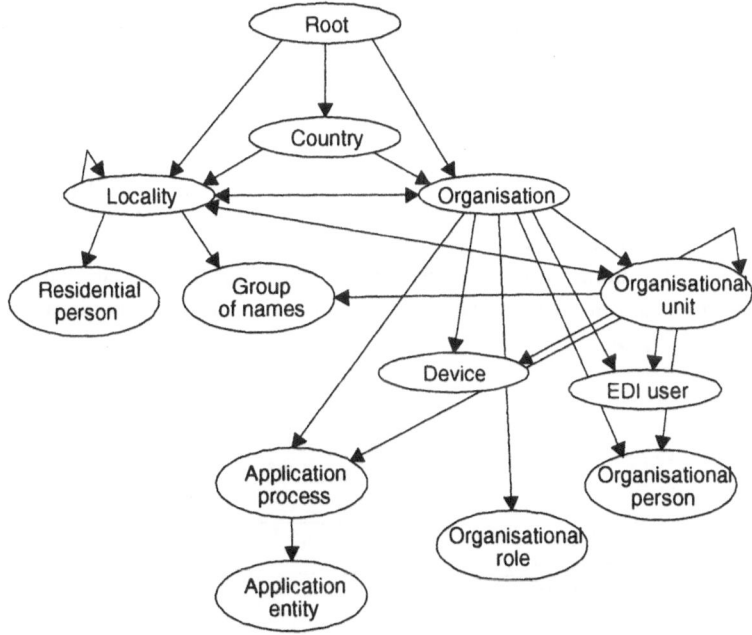

Fig. 13.1. DIT structure for EDI requirements

13.4
Name Resolution

An EDIMG user may use the directory to obtain the message handling O/R address of the EDI-UA corresponding to another EDIMG user. This process is defined as "name resolution" in § 22 of Recommendation X.402.

To obtain the message handling O/R address of an EDI-UA that corresponds to an EDIMG user whose directory name it possesses, an EDIMG user presents the directory name to the directory, and obtains from the directory the attribute message handling O/R address.

To do this successfully, the EDIMG user shall authenticate itself to the directory and have access rights to the information requested.

The directory name may contain a relative distinguished name that is an EDI name. The EDI name can be considered a "user friendly name", as defined in § E.1 of Annex E of Recommendation X.501.

The following figure illustrates an example of an EDI name. Whenever an EDI-UA requires access to an EDIMG user directory entry, using the services of a DUA, it shall construct the distinguished name of the entry. The distinguished name shall contain the EDI name, organisation name of the organisation or naming authority that issued the EDI name, and if required, the country of that organisation. How the EDIMG user obtains the EDI name is a local matter. The EDIMG user shall pass the EDI name to the EDI-UA, who shall pass it to the DUA. When the EDI name is globally unique, the organisation name, and if required, country, may be derived from the qualifier code with the EDI name. When the EDI name is not globally unique, the EDIMG user or EDI-UA shall obtain the organisation name and shall obtain the country, if required, via other means.

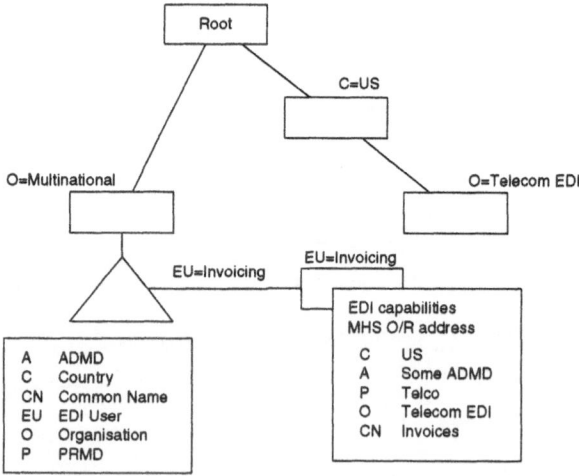

Fig. 13.2. An aliasing example

An alias name may be used to direct the search for a particular entry, for example, to enter the directory with an organisation name and the EDI name in order to extract an message handling O/R address. The figure above shows an EDI-UA identified with the name (O=Multinational, EU=Invoicing). It is also identified by (C=US, O=Telecom EDI, EU=Invoicing). Both EDIMG user names resolve to the same message handling O/R address (C=US, A=SomeADMD, P=Telco, O=Telecom EDI, CN=Invoices).

The following figure illustrates that if the organisation is not an international organisation then the EDIMG user can still be accessed using country as a component of its name.

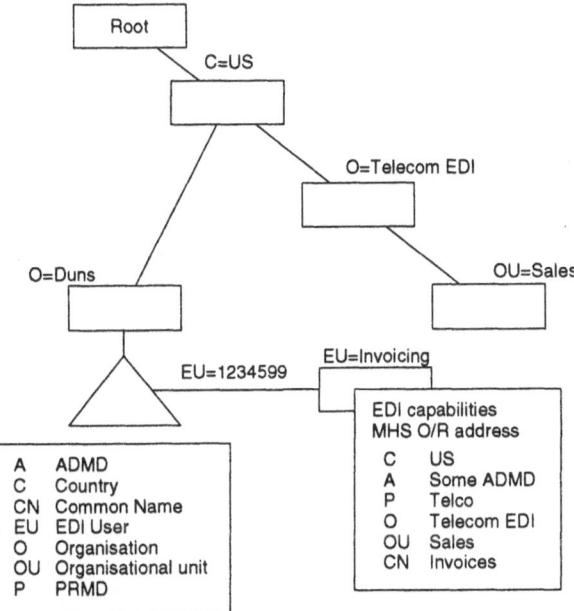

Fig. 13.3. A country-oriented aliasing example

13.5
Authentication

An EDIMG user may accomplish authentication using information stored in the directory. This usage is as defined in Recommendations X.400 and X.509.

13.6
Capabilities Assessment

An EDIMG user may assess the capabilities of another EDIMG user via the directory. Capability assessment allows the EDIMG user to determine, for example, whether the other EDIMG user can process a specific version or the release of an EDI document.

The following directory attributes represent EDI capabilities in the EDI messaging service:

- standard;
- standard version;
- standard syntax identifier;
- document type;
- document version;
- document release;
- controlling agency;
- association assigned code;
- EDI character set.

To assess a particular capability of an EDIMG user whose directory name it possesses, the EDIMG user shall present that name to the directory and request from the directory the attribute EDI capabilities.

To do this successfully, the EDIMG user shall first authenticate itself to the directory, and have access rights to the information requested.

Index

Index 183

MTA ..... XIX, 4-8, 11, 14, 70, 72, 139, 143, 144, 145, 160, 163, 164, 165
Multipurpose Internet Mail Enhancements .... *See* MIME

N

National Institute for Standards and Technology ..... XIX, 129

O

O/R ..... XIX, 6, 7, 9, 12, 14-18, 95, 98, 141, 175, 178, 179
Open Systems Interconnection ... *See* OSI
Originator/Recipient ..... *See* O/R
OSI ..... XIX, 3, 8, 9, 10, 128

P

PEM ..... XIX, 168, 170, 172, 173
Physical Delivery ..... XIX, 10, 20, 141
PKCS ..... XIX, 170
Post Telephone and Telegraph Administration ..... *See* PTT
Privacy Enhanced Mail ..... *See* PEM
Private Management Domain ..... *See* PRMD
PRMD ..... XIX, 6
PSTN ..... XIX, 21
PTT ..... XIX, 5
Public Key Cryptographic Standard ..... *See* PKCS
Public Switched Telephone Network ..... *See* PSTN

R

RFC 1049 73
RFC 1327 ..... 16, 18, 19, 96, 98
RFC 1421 ..... 73, 170
RFC 14221-RFC 1424 ..... *Siehe* PEM
RFC 1487 104
RFC 1496 ..... 18, 19
RFC 1522 61
RFC 1651 ..... 61, 62, 67
RFC 1652 78
RFC 1767 167

RFC 1777 104
RFC 1848 173
RFC 1869 ..... 61, 62, 67
RFC 1891 ..... 68, 71
RFC 1894 ..... 68, 71
RFC 1939 ..... 80, 83, 84, 87, 88
RFC 2015 172
RFC 2045-RFC2049 ..... *See* MIME
RFC 733 ..... 27, 45
RFC 820 44
RFC 821 ..... 27, 47, 60, 62, 64-66, 69, 72, 73, 77, 97
RFC 822 ..... 18, 27, 28, 32-35, 39, 43-46, 49, 56, 61, 63, 66, 67, 70, 72-74, 77, 80, 90, 95-99
RFC1321 91
RFC1734 82
RSA ..... XIX, 151-153, 171, 173, 174

S

Simple Mail Transport Protocol ..... *See* SMTP
SMTP ..... XIX, 1, 2, 18, 36, 47, 48-56, 59-72, 77, 80, 95-97, 128, 149

T

TCP/IP ..... XX, 1
TEDIS ..... XX, 130
Teletex ..... XX, 3, 10, 20
Telex ..... XX, 10, 20
TP ..... XX, 107
TRADACOMS ..... XX, 118, 125
Trade Electronic Data Interchange Systems ..... *See* TEDIS
Trading Partner ..... *See* TP
Transmission Control Protocol/Internet Protocol ..... *See* TCP/IP

U

UA ..... XVIII, XX, 4, 6, 7, 8, 11, 12, 14, 73, 76, 77, 80, 131, 132, 134, 135, 136, 137, 138, 139, 144, 146, 159, 160, 161, 162, 163, 164, 165, 178, 179
UN/ECE ..... XX, 123, 124, 127
UN/EDIFACT ..... XX, 118, 120, 167

Springer
and the
environment

At Springer we firmly believe that an
international science publisher has a
special obligation to the environment,
and our corporate policies consistently
reflect this conviction.
We also expect our business partners –
paper mills, printers, packaging
manufacturers, etc. – to commit
themselves to using materials and
production processes that do not harm
the environment. The paper in this
book is made from low- or no-chlorine
pulp and is acid free, in conformance
with international standards for paper
permanency.